O.J. LINE

OUR
CHANGING
CLIMATE

THE McGRAW-HILL HORIZONS OF SCIENCE SERIES

The Gene Civilization, François Gros.

Life in the Universe, Jean Heidmann.

Our Expanding Universe, Evry Schatzman.

Earthquake Prediction, Haroun Tazieff.

The Future of the Sun, Jean-Claude Pecker.

How the Brain Evolved, Alain Prochiantz.

The Power of Mathematics, Moshé Flato.

ROBERT

KANDEL

OUR

CHANGING

CLIMATE

McGraw-Hill, Inc.

New York St. Louis San Francisco Auckland Bogotá
Caracas Hamburg Lisbon London Madrid
Mexico Milan Montreal New Delhi Paris
San Juan São Paulo Singapore
Sydney Tokyo Toronto

English Language Edition

Translated by Nicholas Hartmann
in collaboration with
The Language Service, Inc.
Poughkeepsie, New York

Typography by AB Typesetting
Poughkeepsie, New York

Library of Congress Cataloging-in-Publication Data

Kandel, Robert S.
  [*Le Devenir des climats*. English]
  Our changing climate/Robert Kandel.
      p. cm. — (The McGraw-Hill *HORIZONS OF SCIENCE* series)
  Translation of: *Le Devenir des climats*.
  Includes bibliographical references.
  ISBN 0-07-033710-1
  1. Climatic changes. I. Title. II. Series.
QC981.8.C5K3613 1992
551.6—dc20   91-29175

The original French language edition of this book
was published as *Le Devenir des climats*, copyright © 1990,
Hachette: Paris, France.
Questions de science series
Series editor: Dominique Lecourt

TABLE OF CONTENTS

Introduction by Dominique Lecourt 7

I. Towards catastrophe? 19
   Is the Earth in danger? 19
   Regularity and variation 22
       The theater of meteorology and climate 22
       Weather and climate 25
       Numerical prediction models 31
   From climates to climate: change or anomaly . . 34
   Our changing climate 38
       Catastrophe 39
       The return of the Ice Age? 40
       Changes in atmospheric composition and
           global warming 41
II. The balance of nature at risk 49
   The greenhouse effect 50
   The ozone question 58
   Is the world getting warmer? 63
   Models 72
   The myth of equilibrium 77
III. How we know what we know 87
   Remembrance of things past 88
   Prediction quantified 92
   A closely watched planet 94
       Instruments 94
       Platforms 97
       People 109
IV. Climate and the political scene 113
   Climatology and politics 113
   Steering the course of our planet 120

Bibliography 125

INTRODUCTION

The question of climate change has recently acquired unprecedented emotional impact. As the risk of nuclear war appears to be receding, it seems that anxiety about an apocalypse unleashed by human folly has now focused on the uncertain future of the Earth's atmosphere. The media are working hard to feed the morbid fascination stirred up by this concern, while politicians try to derive some secondary benefits from it. Listening to some of the catastrophic scenarios that are now being presented as our destiny, it is easy to forget that the question remains unanswered and that it is primarily a scientific one. But science, in this instance more sensible than the wisdom so loftily dispensed by both sides, is neither offering nor insisting on any of the certainties that are being exploited by the calculated pessimism of some, and the calculating humanism of others.

Robert Kandel's book is in no way intended to suppress the very necessary and legitimate vigilance of every citizen; it does aim, however, to put us on our guard against the obfuscations that lie in wait for us. In its conclusion, it asks researchers to stay out of a game in which, ultimately, neither science nor politics stands to gain anything.

The fact that this book was written by an astrophysicist turned climate researcher is in itself a symbol, one which tells a story of its own and highlights the radical novelty of the research in question. Although the original Greek parent of the the word "climate"—meaning "inclination"—already recognized the tilt of the Earth with respect to the Sun, it took centuries of time and several intellectual, social, and technical revolutions before astronomy could be successfully integrated with meteorology.

When Aristotle wrote his little treatise *Meteorologica* in the 4th century BC, he followed established tradition and subsumed under that term all the transitory phenomena which occur around the Earth: not just atmospheric phenomena in the strict sense (snow, hail, wind, rain), but also phenomena that we would call astronomical, such as the nature of comets and the Milky Way, together with others, such as the origin of the oceans and rivers or the distribution of the continents, which we would group under the heading of "Earth sciences."

Although like most of his predecessors (especially Democritus in the *Tabula Astronomica*), Aristotle gives the Sun a dominant role in the origin of these phenomena, he very clearly sets it in the general context of his cosmology. This cosmology, however, assigned to the stars a quasi-divine status corresponding to the perfect circular motion that they were thought to possess. His theory thus opened up a metaphysical abyss between the "upper" world where the stars resided and the "lower," or sublunar, world. How was it then possible to account for phenomena occurring in the region located between the Earth and the Moon The explanation given by Aristotle rests on the doctrine, borrowed from Heraclitus (576–480 BC), of "exhalation." He proposed the existence of two types of exhalation produced by the action of the Sun's rays on the Earth's surface. When these rays fell on dry ground, they caused it to emit a hot, dry exhalation, which Aristotle compared to smoke but also to fire and wind. When they fell on water, they drew from it a vaporous exhalation. The dry exhalation consisted of particles in the process of turning into fire, while the wet exhalation contained particles of water becoming air; but the latter exhibited primarily the qualities of water, being cold and wet. The upper part of the atmosphere was therefore considered to be filled with dry exhalation, while the lower part was assumed to contain both exhalations and thus to possess the heat of one and the moisture of the other. With dazzling ingenuity, Aristotle then explains the phenomena of the upper region—shooting stars, the Milky Way, comets—and then the phenomena of the lower region arising from the wet exhalation: rain, clouds and fog, dew and frost, snow and hail. Lastly, he addresses the results of the dry exhalation in the same region: wind, thunder, lightning, storms, and thunderbolts. Moreover, using the same basis, he offers his answer to some of the questions that haunted the Ancients: why does the ocean not overflow, given the number of rivers that continually flow into it? Why is it salty? His explanation of the rainbow is still justly famous.

For centuries meteorology would remain dominated by Aristotle's theories, scrupulously summarized and commented upon by Thomas Aquinas, and rigorously separated from astronomy and its figures and calculations.

Two major events in the history of humankind helped us expand our horizons and change our theoretical framework. They are obvious, but worth mentioning: Until the Renaissance, observations concerning climate, while numerous, were confined to the Mediterranean basin and to Europe. Only with the great expeditions of the late 15th-century explorers—Christopher Columbus and Vasco da Gama, to name just two—did Europeans become familiar with very different climatic conditions, and discover the climate of the planet's tropical and equatorial regions. In the 16th century, scholars extracted from these observations, first of all the theory of a regular distribution of temperatures by zones (that is, in bands between two latitudes), and then the idea that these various zones were symmetrically distributed with reference to the equator.

The *Encyclopedia* of Diderot and D'Alembert still defined "climate" as follows: "Portion or zone of the Earth's surface defined by two circles parallel to the equator (...) the climates thus proceed from the equator to the poles, forming as many zones parallel to the equator." The article concludes with the specific information that "one climate differs from the one nearest to it only in that the longest day of summer is a half hour longer or shorter in the one than in the other." The complexity of the problem is alluded to, however: "One must not conclude that the temperature is the same in countries situated under the same climate," since circumstances "are complicated" by the action of the Sun: wind, volcanoes, proximity of the sea, location of mountains, etc.

It was the obvious failings of Aristotelian cosmology, followed by the demolition—at the hands of Copernicus, Galileo, Descartes, and several others—of the physics that underlay it, that revolutionized the context of the problem: from that point on, meteorology could no longer rightfully be separated from astronomy. At the end of the 17th century, the famous English mathematician and astronomer Edmund Halley (1656–1742) explained the origin of trade winds and monsoons by hypothesizing a gigantic convective mixing process in the atmosphere, produced by the Sun. John Hadley (1682–1744), the first person to produce a practical sextant, associated the rotational motion of the winds with the Earth's rotation around its axis. We should add that these theories would never have seen the light of day without expanded use of the barometer, the result of Evangelista Torricelli's famous investigations (1608) of atmospheric pressure. But it was not until the mid-19th century that a decisive addition to these theories was made by the French mathematician and engineer Gaspard-Gustave Coriolis (1792–1843). Beginning with a problem in pure kinematics posed a century earlier by Alexis-Claude Clairaut, he showed that terrestrial mechanics—the science of motion using the Earth as frame of reference—had to be fundamentally revised to take into account the "composite centrifugal forces" which result from the Earth's motion. These forces have been referred to ever since as "Coriolis forces."

The satellite observations and colossal mainframe computer calculations performed today in an effort to understand the dynamics of climate are an extension of this research, in which astronomers, physicists, and meteorologists join forces. This work may be considered the culmination of the vast research program proposed by

Alexander von Humboldt (1769–1859), the true founder of modern climatology, who proposed—in his masterpiece entitled *Kosmos*, the first volume of which appeared in 1845—an initial explanation of the relationship between the oceans and currents and the submerged land, and made that relationship an essential element in his "physical description" of climatic change.

But the idea of a general dynamic of climate has been joined by that of a natural evolution in the planet's climate as a whole. The study of past climates has become an entirely separate discipline called paleoclimatology. The universally recognized father of this science is the Swiss geologist and paleontologist Louis Agassiz (1807–1873), who demonstrated in 1837 that certain striated rocks found in the Jura were vestiges of the Ice Age that is believed to have preceded our own epoch. Numerous similar observations in Northern Europe finally yielded convincing proof that glaciers had in fact covered Europe from the 50th to the 72nd parallel, a distance of 2,300 kilometers [1,400 miles]! Almost immediately, an effort was undertaken to date this phenomenon of glaciation. The calculations of James Croll (1821–1890) yielded a first approximation, but it was not until very recently, in the 1950s, that the carbon-14 dating method invented by the American chemist Willard Frank Libby allowed scientists to determine the age of the great glacial periods which, in the meantime, had been found to consist of more than just a single episode. The date of the maximum extent of the last glaciation was set at 18,000 years ago.

Other investigations revealed yet more pages of the planet's climatological archives; in particular, the study of changes in sea level and of marine sediments which had already been suggested by J. Croll. But archeology and paleontology also lent a hand. For example, André Leroy-Gourhan's brilliant work at the site of Pincevent uncovered fossil reindeer bones in the Seine valley, establishing that 10,000 years ago the Paris region had a climate similar to that of Lapland!

The idea that, appearances to the contrary notwithstanding, our climate might have undergone considerable variation over the course of history has therefore gradually become established over the last two centuries. A consistent network of proof has resulted in calculations that are now extremely reliable.

All that remained was to establish the relationship, sought since the mid-19th century, between climatic cycles and astronomical phenomena. During the 1930s, this was the goal of a Serbian mathematician named Milutin Milanković, a relative of Alfred Wegener who was at that same time promulgating the theory of "continental drift." On the basis of an immense study of the Earth's insolation periods, covering 600,000 years at various latitudes, Milanković uncovered the role played by the inclination of the Earth's axis with respect to its orbital plane, which varies between 22° and 25° with a period of 41,000 years. Milanković's theory, like that of Wegener, was not well received by the scientific community and lay forgotten for more than twenty years. Observations have now restored it to a place of honor.

But although there are also climatic cycles on a scale of tens and hundreds of thousands of years, and although these cycles suggest that an Ice Age will return (a period of 50,000 years is being suggested), we have found that human activity has just introduced, on a completely different scale, a perturbation which is particularly frightening because the magnitude of its effects is still difficult to evaluate. Here is where climatology collides with politics, and the interaction these days is very strong. Robert Kandel gives a very vivid account of the confused debate surrounding the climatologists "modeling" process. A philosopher might be tempted to hark back to an earlier and celebrated encounter between "climate theory" and political thought, as a way of casting some light on the matter from his point of view. Montesquieu certainly would never have defended the climatic fatalism that is sometimes attributed to him, and the Encyclopedists were correct to emphasize that the famous Chapter 14 of *L'Esprit des Lois* [The Spirit of the Laws] does not make "everything depend on climate." He simply points out its influence on the customs, character, and laws of various peoples. But can we ignore the harsh fate reserved for inhabitants of the "hot countries," consigned to rigorous laws and exemplary punishments because of the putative "softness" of their temperament resulting from the heat? Montesquieu indeed condemns slavery with unforgettable indignation, but some expansionist politicians might find in his arguments the justification for some decidedly less innocent practices.

The new encounter between climate research and politics that is taking place before our eyes is occurring in an entirely different context. It is being described as an indication, not of the constraints being placed by Nature on relationships among human beings, but of the excessive power acquired by our kind over a Nature that has been denatured by being incessantly humanized. Nevertheless, it is not amiss to wonder whether one of the essential elements in today's loudly trumpeted planetary catastrophe scenarios should not, once again, be sought in the "hot countries." Does not climatic fatalism have the effect of sidestepping the very socioeconomic terms under which the immense question of Third World development is being addressed, thus pointing us, silently and with scientific justification, towards a solution that merely prolongs the current trend—namely, aggravated underdevelopment? This suspicion is by no means the least interesting idea discussed in Robert Kandel's reflections.

Dominique LECOURT

To the memory of my father,
Max Kandel.

I

TOWARDS

CATASTROPHE?

IS THE EARTH IN DANGER?

The Earth, we are told, is in danger. The threat is depicted as global and radical: our entire planet is courting catastrophe. For some time, there has been an apocalyptic accent to discussions about "climatic change." Where once all was unvarying stability, what is emphasized now is the potential fragility of the biosphere—the entirety of the animal and vegetable kingdoms—in the face of increased human activity. Building progress in space exploration and expanding their satellite observations, specialists are asking what makes our planet unique. At the same time paleoclimatology, the study of past climates, is advancing by leaps and bounds, opening up to us the historical, glacial, and sedimentary archives of the planet and revealing natural changes on a variety of time scales.

Questions and hypotheses about the causes of these changes are continually arising, and one question is being asked with particular urgency: in what ways are people affecting the planet's climate? Using computerized numerical "models," can we determine the future effect (or "climatic impact") of these actions?

19

Politicians, who for years have ignored the warnings of scientists and environmental protection movements, are suddenly becoming alarmed and invoking public opinion to bear witness. They are asking researchers for estimates, and above all for certainties; they are announcing and making decisions on an international scale. The media, naturally inclined toward the spectacular, are distilling from studies of the ozone layer, or on the greenhouse effect, the most emotionally charged issue: the universal, naked threat. A new ideology is winning minds, based on a rationalization of the old nightmares whose ability to fascinate and terrify has been experienced by religions for centuries: it deplores the irremediable corruption of nature by human actions and announces that the world will end tomorrow. Amid these confused rumors it has become difficult to see clearly. Very often, the important economic, social, and human questions that constitute obstacles to the researchers' forecasts are simply dismissed and science finds itself called upon to give its blessing to this obfuscation.

In the pages which follow, I intend to leave aside for a moment the catastrophism that is presently in fashion and to address the problem of our changing climate as a scientist and a researcher, that is, with emphasis on unanswered questions rather than certainties. Climate research owes its progress to the contributions of a wide variety of disciplines: meteorology and climatology of course, but also glaciology, oceanography, astronomy, botany, paleontology and many others. We will see how refinements in our observation, dating, and calculation methods are playing a decisive role. Naturally I cannot claim to be an expert in all these fields and, despite my efforts to give an objective overall view, what I am presenting in this book may perhaps still be tinged with my personal preoccupations resulting from my training as an astrophysicist specializing in the interactions between radiation and gases who has become "converted" to the observation of the Earth and its atmosphere from space after many years of studying stellar and solar atmospheres from Earth.

Restricting this book to the specialized field of my research would have meant presenting a partial and incorrect view of this eminently interdisciplinary science that climate science has become. But in attempting to present this subject to the non-specialist reader, I feel an obligation to make some occasionally gross simplifications and, despite all my efforts, I may have let some errors slip into the text. I hope the specialists will forgive me. I am also running other risks, since everybody has something to say about the weather. I make no claim other than to explain some aspects of the Earth, a planet with which all my readers are very familiar in some detail, at least as far as some part of it is concerned. It is not like the planet Neptune, the Orion Nebula, or the Andromeda galaxy, about which non-specialists, unless they are amateur astronomers, will gladly accept what the astronomers tell them without really having any way of contradicting what they are told.

As a researcher, I am of course fascinated by the great global experiment to which our planet is being subjected by human activities. As a scientist, I owe it to myself to remember the uncertainties in what we know, and I will undoubtedly betray my irritation at the exaggerations and even untruths that are making the rounds. But at the same time, as an inhabitant of this planet, I cannot remain an unconcerned spectator when faced with disturbing developments in certain points of view. I am aware that, in the political and economic world where self-styled "decision-makers" reside, it would be desirable to act with some knowledge of the consequences of one's actions; moreover, every decision, even to do nothing, produces its own consequences, and "fact files" are never perfect. However, although scientists have a duty to respond to the appeals of the political world, they cannot pretend to accept as a given something that is still uncertain. I hope that when you have finished this book, you will understand more clearly where, at the present state of our knowledge, the domain of "scientific questions" ends and that of "socio-political questions"—where all that counts is the citizen's judgment—begins. But, the citizen must be well-informed…

REGULARITY AND VARIATION

The theater of meteorology and climate
Before beginning our inquiry, let us set the stage: meteorological phenomena take place essentially in the atmosphere, but the processes which unfold there affect those which occur in the oceans and on the surface of continents and ice caps, and are in turn affected by them. These processes depend on the Sun, but have no influence on it.

The first of these processes is the conversion and transport of solar energy. At the average distance between the Earth and Sun, the solar energy radiation flux is approximately 1,368 watts per square meter (W/m^2), essentially in the form of visible and near-infrared radiation (up to a wavelength of 4 micrometers). This is equivalent to an average flux of 342 W/m^2 on the "top" of the atmosphere, namely at an altitude of a few dozen miles, bearing in mind that it is always night over half the Earth's surface. The heat flow emerging from inside the Earth (0.06 W/m^2) is negligible by comparison. The Earth and its atmosphere reflect on average 30% of the solar flux; clouds (as well as light-colored or snow-covered surfaces, and deserts) are responsible for most of this reflection. We are therefore left with an average flux of about 240 W/m^2 which is absorbed and converted into heat. It is important to note that absorption occurs primarily at the surface; in other words, the atmosphere is essentially heated from below, which may seem paradoxical since the source of this energy is the Sun. Following this conversion of radiation to heat, this energy is finally re-emitted into space in the form of infrared radiation (at wavelengths between 40 and 50 micrometers). If this global radiation balance, that is, the difference between absorbed solar radiation and emitted thermal radiation, is not zero, the temperature of the Earth–atmosphere system must change.

We can see immediately that, locally and at any given time, the radiation balance is almost never zero. In equatorial and tropical regions it is positive, meaning that there is an excess of absorbed solar energy compared with the infrared flux emitted into space; the same is true in summer at middle latitudes (Europe or the United States). Of course the opposite is true in winter or in polar regions. This phenomenon is possible because the atmosphere and the oceans transport heat from the "excess" zones to the "deficit" zones. Without this heat transport, temperature differences as a function of latitude and season would be much more severe than those we are familiar with. This redistribution of the highly unequal distribution of solar energy is the first essential function of what we can call the climatic system.

"All the rivers run into the sea; yet the sea is not full; to the place from which the rivers come, there they return again." This verse from Ecclesiastes (1:7) reminds us of the second essential function of the climatic system, namely to maintain the hydrologic cycle and irrigate the Earth. This cycle is closely linked to the energy cycle, since energy is needed for water to evaporate, and it is the solar energy flux which supplies this energy. The solar flux is especially strong on subtropical seas, where there is little cloud cover. Water evaporates and is carried away by winds; the latent heat required for evaporation is given back to the atmosphere when the water vapor condenses into the liquid droplets or ice crystals that form clouds. Outside the tropics, for instance along the Gulf Stream, water which has been heated in the tropical and subtropical regions is transported to higher latitudes where it returns its heat to the air passing over it. This is why Brittany enjoys a much milder climate than Newfoundland, although they are at the same latitude.

The climatic system is therefore based on movement, especially that of winds (which originate from differences in temperature and pressure) and of ocean currents, which are in fact set in motion by the winds but which, because they themselves transport heat, in turn modify the conditions which give rise to the winds. Dominated by the distribution of solar radiation and by the Coriolis force resulting from rotation of the Earth—which transmits some of that rotation to any moving body of fluid—the global circulation of the atmosphere and the oceans has certain regularities, such as a large-scale organization which follows the seasonal cycle. This circulation also comprises smaller and more variable structures, such as the atmospheric eddies that constitute low-pressure systems and the rings of warm or cold water that can be seen by satellite in the ocean. These structures play a primary role in transporting heat and water. They constitute the field of meteorology in the strict sense, which requires measurements of pressure, temperature and humidity, and wind strength, but also measurements of solar and infrared radiation, and of the content and macro- and microscopic structure of clouds. The difficulties encountered in understanding these phenomena have nothing to do with the laws that govern them, which are the laws of conservation of mass and energy and Newton's laws of motion; rather, they arise from the multiplicity and complexity of the interactions involved.

Weather and climate
No one will argue that it is difficult to predict changes in the weather. Meteorologists pay the price for that uncertainty

every day, and public opinion never fails to poke fun at their mistakes. But the public is unfair, since it does not greet the forecaster's unarguable successes with the same enthusiasm; despite the fact that satellite images are shown on television every night, it has not truly comprehended the difficulties involved and the considerable progress that this discipline has managed to make. Focussing on the foul-ups of weather forecasts, it is easy to forget the exact definition of a climate: a certain regularity underlying apparent disorder. We nevertheless have a fairly precise idea of what is meant when we compare the climate of Africa with that of Europe: one has certain regularities that are not found in the other. But of course regularity does not mean either consistency or even perfect periodicity.

For example, we might establish an annual cycle averaged over one decade, and this would adequately describe the characteristics of the seasons during the year. But variations may occur from one year to the next: at a given location, cold years may be followed by hot years, or dry years by wet years. In Western Europe, 1989 repeated to a certain extent the extreme dryness of the summer of 1976. The glacial cold of the winter of 1962–63, like that of 1986–87, was followed by two fairly mild winters. During the summer of 1988, extreme dryness and heat occurred over much of the United States, while the summer of 1989 brought lots of rain.

In certain regions of the world, much more so than in Europe, there is an even more obvious alternation between "fat" years and "lean" years; this is especially true for semi-arid regions such as the Sahel where a series of very dry years will follow a series of years in which precipitation is (relatively) abundant. The regularity which defines the climate therefore includes variability, and in many places a year in which the temperature faithfully followed the average annual cycle, with no deviation, would be a very abnormal year. Technically, we can speak of the "time spectrum of the variations": astronomical conditions impose certain diurnal and annual periodicities, but the variability which characterizes the combined system of Earth, oceans and atmosphere reveals certain quasi-periodicities ranging from a few days to at least ten years, with a particular time scale for each regional climate.

Average values—such as average monthly temperatures and precipitation levels—are not sufficient for complete characterization of a climate. We must also know the extremes on either side of these averages, and the variety of situations that can be found within them. These observations are of enormous practical importance: ignoring the one cold winter that occurs every ten years on average (or better, ten times each century) is asking for trouble; ignoring the once-a-century record cold temperature may turn out to be an acceptable risk in some cases but not in others. For example, we might accept total paralysis of transportation for a few days, once or twice a century, but since the catastrophic floods of 1953, the Dutch have used a "hundred-year storm" as the basis for designing their coastal defenses.

In general, we speak of a "climatic anomaly" when confronted with an extreme situation involving temperature and/or precipitation over a period of at least ten days.

The question is then to ascertain which time scale will allow us to determine the reference average, and the percentage of situations that must be considered extreme. For a given region, the annual mean established over a thirty-year sample period, and the deviation of each annual mean from this climatological average, can be of some help, but it may be much more useful to calculate averages by season or by month. Determining the "abnormality" of a given month—for example July 1989 in Southern France—comes down to determining the difference between the mean and the average of the ten or thirty preceding Julys, and how this deviation compares to the average deviation.

We can therefore see that, in general, the notion of "average" must be handled with care, since sometimes the average situation is no more common than a situation referred to as "extreme." This is the case, for example, in Winnipeg, on the Great Plains of Canada, where the average temperature in January, calculated over several decades, is –19°C [–2.2°F], but where the average for a given month of January is just as likely to be –12 or –25°C [+10.4 or –13.0°F]. If we examine what in fact happens from day to day, we find either very cold, very dry weather with no clouds, with temperatures that can go down to –30°C [–22.0°F] or even –40°C [–40°F], or wet, cloudy, and relatively mild weather. If we take the average of these two situations, we come up with a value that is almost never encountered in real life. In this case, therefore, the climate is defined much less by the average and much more by the "see-saw" between the two opposite situations, and by their relative frequencies.

The ten-day period (called in French a *décade*) mentioned earlier as the minimum for climate studies is a fairly arbitrary choice. The climate on New Year's Day at a particular place may certainly be of interest in determining what kinds of festivities might reasonably be organized there. However, there are very few significant climatological differences between that day and Christmas, or between July 4 and July 14. Climatological information obtained by compiling statistics over a fairly long period (say 10 to 30 years) is available well in advance. The problem of weather forecasting is a very different story; information about the climate in Paris in mid-July has been available for years, but a prediction of the weather for the Bicentennial Bastille Day parade on July 14, 1989, was not reasonably possible before July 1 of that year.

Forecasting what the weather will be within several days, or climatological prediction of the probabilities of a particular type of weather—the choice will depend on social and economic expectations. For agricultural needs, the emphasis will fall essentially on the amount of rain that can be expected during the spring or summer months. In Europe, the fact that it rains on one day rather than another is fairly insignificant, except, of course, when choosing a day for haymaking. But in semi-arid regions, a prediction of the particular ten-day period during which the first rains of the wet season will arrive—a prediction that is at the limit of climatology—can be essential in deciding on the date to sow. This definitely constitutes an urgent demand from farmers, a demand to which agrometeorology is working hard to respond, and with considerable difficulty.

In regions where winters are cold, weather prediction plays an important role in decisions affecting fuel supplies or electricity generation. Heating requirements, on the other hand, which determine the choice of a particular type of construction, depend on climatology. But because people often hesitate to invest on the basis of extreme situations, which are certainly less frequent but still part of the climate, houses in Mediterranean countries are often uncomfortably cold in winter. As far as maritime navigation is concerned, in Antiquity it required nothing more than general climatological information combined with the highly empirical art of examining the sky to decide what the weather would be. That was how routes were selected or avoided. Today, of course, we ask much more of meteorology, whether for a giant oil tanker or for a sailboat making a solo trip.

With contemporary means of transportation, and especially in aviation, a need for short-term predictions has arisen: meteorology has thus become the subject of much more stringent, and very different, requirements. Today we use the word "nowcasting" rather than forecasting to describe the art of immediate prediction: the pilot of an aircraft about to cross the Atlantic needs to know the situation for a journey that will take only a few hours. And what is true for aviation also applies to many other activities of modern life. Under these conditions, the task of the meteorological services is first of all to collect, without delay (that is, in "real time"), the most complete information possible about the situation as it exists over the entire globe (that is, obtaining a "synoptic" view), then to analyze this information to eliminate the inevitable incorrect data, use the analyzed information in short-term forecasts (from a few hours to a few days), and finally to distribute the results of these analyses and predictions with a minimum of delay.

Numerical prediction models

A climate could be defined as a synthesis of meteorological variations. The "models" used to study climate involve a more or less simplified representation of the components of the climate system and the processes which govern its status and evolution. The most elaborate and complete of these models are derived from the "general circulation models" used in deterministic numerical weather prediction. The concept of numerical weather prediction, originated by the Englishman Lewis Fry Richardson (1882–1953), dates from 1922. The first test, performed with the assistance of a small army of (human) computers (an extinct profession), required months of work for a twenty-four hour prediction—and the results were wrong. It was only with the development of electronic computing methods at the end of the 1940s that promising results were obtained. Today, all the major meteorological services "run" these models through their high-powered computing centers. The most advanced of these models is used at the European Center for Medium-Range Weather Forecasting at Reading, in England.

What exactly is involved? The problem is to predict motions and physical changes in the atmosphere and on the Earth's surface based on a necessarily imperfect knowledge of an initial state defined by the distribution—over the globe and as a function of altitude—of pressure, temperatures, humidity, wind, etc. at a given moment. It is possible to write equations that describe the motions and physical state of air masses at each point and each instant. In order to proceed along practical lines, however, some spatial carving up must be applied, delimiting the boundries of regions several hundred kilometers square and dividing the atmosphere into ten or twenty horizontal layers. This slicing then forces us to represent on a rather coarse scale processes which are occurring at smaller spatial scales. For example, it is obvious that using a single number to describe the clouds present inside an atmospheric cell 250 km square [150 miles × 150 miles] and 1 km [3,300 feet] thick can be no more than a caricature.

Equations then determine the effects on other cells of changes in one of them. The whole system is then represented by a very large set of calculation programs, which requires enormously powerful computers to predict changes in the weather fast enough for the results to arrive before the facts. These calculations are restarted at regular intervals (generally every 12 hours), by introducing and "assimilating" new data; this prevents the model from drifting too far away from reality, at least as far as the shortest range predictions are concerned. At present, the results are relatively good in predicting the weather a few days in advance, and efforts are being made to extend that to about ten days. Beyond two weeks, determinist numerical prediction appears unachievable, since what will happen then is sensitive to details neglected and unknown at the initial state, which cannot be determined. However, even if such models cannot predict the weather beyond two weeks, it is believed that, with a certain number of adjustments, they can tell us about what the climate might be in a situation different from the one we are familiar with today.

Although the English language makes a clear distinction between "time" and "weather," French and the other Romance languages gloss over the ambiguity between chronometric measurements (*temps* = time) and barometric values (*temps* = weather). This does nothing to prevent bad weather in the Mediterranean! Astronomical time has in fact served for many years to define chronological time. Astronomers can predict eclipses to within a second, years in advance, and planetary motions for millions of years. Why is it that meteorologists cannot do the same, and why is the weather characterized by such disorder? As we have just seen, meteorology involves the movement and transport of energy and water in the atmosphere; even with the spatial division described earlier, which constitutes a gross simplification, we must study hundreds of thousands of atmospheric "cells" that obey the laws of gravitation and fluid dynamics, each undergoing numerous interactions— mechanical, thermal, radiative, and even chemical—with the others, and none playing a truly predominant role. The motions of the solar system, on the other hand, depend practically exclusively on the law of universal gravitation and Newton's three laws of motion, linking about a dozen massive bodies that are essentially point-like compared to the distances that separate them, one of which (the Sun) dominates the system from afar, while Jupiter dominates the other planets. Once we have a good understanding of an initial state (the mass, position, and velocity of these objects at a given moment), we can then realize Laplace's dream and predict their future motions for a very long period—although not for eternity. When astronomers attack a problem as complex as terrestrial meteorology—for example the prediction of solar flares—they run into enormous difficulties. And yet the Sun, an entirely gaseous body dominated by hydrogen and helium, where not a single complex molecule exists, still seems quite simple compared to the atmosphere of Earth, with its clouds of liquid and solid water surrounding oceans and continents unequally distributed between the hemispheres, all of them variously heated by the Sun.

FROM CLIMATES TO CLIMATE: CHANGE OR ANOMALLY?

Returning to the question of climate, climatological services have traditionally defined it on the basis of meteorological parameters averaged over thirty years. If we go along with that definition, it is difficult to talk about climatic change on a shorter time scale. There is a new factor, however: the accelerated development of human population and human economic activity is leading to significant modifications in parameters such as atmospheric composition and vegetation cover, on shorter and shorter time scales. Waiting thirty years to determine whether climatic change has occurred may therefore appear too long, if what we want is to understand these changes and decide on what measures to take to prevent or perhaps correct their effects. It has therefore been suggested that ten-year periods be used to define a climatological approach.

This immediately raises another problem: does the observation of a difference between one decade and the next constitute a real change in climate, or is it instead an anomaly—a fluctuation—within the same climate? The question is particularly relevant to the Sahel, the southern "shore" of the Sahara, a transitional zone (now located at about 16° North latitude) between a very arid desert to the north and forested areas farther south. There is no doubt that major climatic changes have occurred in West Africa. At the maximum extent of the last glaciation, 18,000 years ago, the Sahara extended hundreds of miles farther south, and fossil dunes can be found in what is now the Sahel; 10,000 years later, Sahelian vegetation had advanced to 21°N. A portion of today's Sahara was therefore green, and the vegetation has since receded. But we must not confuse these large-scale trends with the more rapid fluctuations revealed by various historical and geological studies.

Has a change occurred in the Sahel's climate since 1970? It is true that the average precipitation observed since that date is definitely lower than in the 1950s and 1960s. Particularly severe droughts occurred in 1972 and 1973, then in 1983 and 1984, with dramatic human consequences that were brought to world attention by television and the press. The victims of the 1941 drought—which was just as bad—got no such attention. Was this a climatic change? Some people have not hesitated to trumpet the fact. To be sure, sudden changes have been detected in the past, especially the disappearance of the forests at 16°N in

West Africa about 2,000 years ago. But since 1985 the rains have been more abundant, sometimes even excessive. Although the word "anomaly" is too weak to describe this cruel episode which lasted some twenty years, it still did not involve a climatic change. The term "climatic crisis," proposed by Pierre Rognon of the Université Pierre-et-Marie-Curie in Paris, seems to be a good description of this phenomenon, which must be understood as an intrinsic element of the Sahelian climate.

Another interesting and spectacular phenomenon that until recently had remained highly enigmatic goes by the name of "El Niño" (the Christ Child), since it occurs around Christmas on the Pacific coast of South America. A cold current flowing northward off the coast of Peru was discovered many years ago by Alexander von Humboldt (1769–1859), in whose honor it was named the "Humboldt Current." Although the coast is an arid region (containing the Atacama desert, one of the driest in the world), the sea is extremely fertile since it is supplied with nutrients by upwelling cold water. These elements support a long food chain that begins with phytoplankton and progresses through fish, birds, mammals (think of the beef cattle fed on anchovy meal), and humans. Fairly regularly, around Christmas, the wind changes direction; the current and cold water upwellings become weaker, and a small warm current arrives: the "Corriente del Niño." Occasionally this warm current becomes established, and the change in wind direction persists. The temperature of the ocean's surface layers then rises above its mean—by 1, 2, even 5°C [1.8, 3.6, 9°F]—and an "El Niño" is then considered to be present. This phenomenon occurs every four, five, or seven years, with no periodic regularity, although it can be relied on to happen twice every decade.

The "El Niño" phenomenon obviously has a major effect on the local economy, particularly because it determines whether anchovy fishing conditions are favorable or unfavorable. But about fifteen years ago, we began to realize that it forms part of a process on an even vaster scale, which affects the entire southern and equatorial Pacific and is in turn linked to what is called the Southern Oscillation. This phenomenon, first described in 1924 by Gilbert Walker, an Englishman working for the Meteorological Service of India, consists of an enormous see-saw between the eastern and western sides of the southern and equatorial Pacific: by comparing sea-level atmospheric pressure readings at Jakarta, in Indonesia, with those on Easter Island, it was found that when the pressure is higher than average in the west, it is low in the east, and vice versa. This see-sawing also affects winds, the level of the sea itself, and the temperatures on the surface of the ocean. Years of low pressure in the east coincide with the appearance of "El Niño." Researchers now refer to "ENSO events" (El Niño/Southern Oscillation), the effects of which extend over the *entire* equatorial Pacific basin (and, according to some, as far as India and Africa).

Ordinarily, a great deal of precipitation falls on Indonesia and Northern Australia, while there are few clouds and almost no rain over the tropical latitudes of the central and eastern Pacific. During "El Niño" years, however, Indonesia suffers from a relative drought. The extreme drought which plagued the region during the 1982–1983 ENSO event was certainly not unrelated to the fire which then ravaged a region the size of Switzerland on the island of Borneo. But each ENSO event brings a great deal of rain to the islands farther east, which then unleashes a sudden and spectacular change in the vegetation. On Tahiti, there were five typhoons in a row during the 1982–1983 ENSO event. In extreme cases, there may even be rain in the desert regions of South America.

We can thus state that this irregular phenomenon corresponds to a climatic "anomaly," since it gives rise to a situation radically different from the average situation that is referred to as "normal." But it is obvious that this irregularity occurs... fairly regularly! It should therefore be integrated into the very definition of the climates of these regions. This has in fact been done by most climatologists who are learning more about this process, and are beginning to be able to predict its onset.

OUR CHANGING CLIMATE

The idea that there has been some evolution in the Earth's climate as a whole and that climatic changes have occurred, although it is not new, has been confirmed and clarified since the last century. We now know that there has indeed been a major change in climate since the arrival of people on Earth, and that 18,000 years ago, at the height of the Ice Age, the climate was radically different from what it is today. We also know that small climatic fluctuations have occurred on the scale of a few centuries: Emmanuel Le Roy

Ladurie's book *Times of Feast, Times of Famine: A History of Climate since the Year 1000* demonstrates this very well. We can also see, by comparing recent photographs with 18th- and 19th-century engravings, that certain glaciers (such as the Argentière glacier in the Chamonix valley) have visibly retreated.

But the reason why the evolution of our planet's climate is a particularly acute question right now, and why it is so fascinating and worrying, is that the growth of human population and human technology is beginning to produce very perceptible effects, at a faster and faster pace, on the atmosphere surrounding the Earth. In short, we are modifying the physical state of our planet on a global scale, in a manner which can no longer be ignored.

Catastrophe

Are we threatened by catastrophe? Like much of what we read in the newspapers, the word may seem excessive in terms of the actual changes involved. Real catastrophes are always frightening—and there have been some, if we believe those who attribute the extinction of the dinosaurs, at the end of the Cretaceous era some 65 million years ago, to the impact of a large asteroid or a comet. Have shocks of this kind, both climatic and ecological, played a major role in the evolution of species? Some believe so. Could the same thing happen again tomorrow? We cannot rule it out, but it could just as well happen 300 million years from now! Without diverting attention to a discussion of the meaning of life in a Universe where the "end of the world" might arrive any day, I must admit that I have not lost any sleep over this eventuality, and I certainly would not recommend installing a warning system.

A few years ago, a great deal of ink was spilled concerning the possibility of another climatic catastrophe, in this case "nuclear winter." The idea was that if the smoke from the cities incinerated in a nuclear war spread around the planet and remained in the atmosphere for long periods, thus blocking solar radiation, surface temperatures would drop. A prolonged decrease would certainly constitute a climatic and ecological catastrophe, especially for tropical species which are not adapted to the cold. In fact, research undertaken in the United States, the Soviet Union, and elsewhere, has yielded much less extreme results: it would be more accurate to talk about a "nuclear autumn." I confess that I was rather uncomfortable with the controversy surrounding this question: whatever the climatic consequences of a nuclear war, the catastrophe would be primarily human and moral, and would constitute the most monstrous failure of our civilization.

The return of the Ice Age?
For two million years we have been living in a period when ice sheets alternately advance and recede: as we have seen, the last glacial maximum period occurred in the fairly recent past, 18,000 years ago. Although from a certain point of view the changes had only a limited scope—after all, the ice did not cover the planet's entire surface—there is no question that some very large-scale changes did result. Sea level was more than 100 meters lower because of the volume of water locked up in the ice caps which covered

much of Europe and North America, and the Sahara was much more extensive than it is today because of changes in atmospheric circulation. These changes follow a semi-regular rhythm: quasi-periods of 20,000, 40,000 and 100,000 years. For several years, scientists have been taking seriously the calculations of the Serbian researcher Milutin Milanković, who, on the basis of earlier ideas, deduced the ebb and flow of the glaciers from variations in the Earth's orbit and rotation, and therefore in the distribution and seasonal cycle of solar radiation on the planet's surface. These variations continue and, if no new factors come into play, we can expect the ice to return, probably again covering Canada and Scandinavia at the very least. It will not happen tomorrow, but perhaps the day after tomorrow (say, 70,000 years from now), when we have some right to hope that humanity will still be around.

Changes in atmospheric composition and global warming

People have been influencing the biosphere for at least 8,000 years, since the invention of agriculture. But for several decades, it is the very composition of the global atmosphere that they have been modifying. To start with the most obvious and perhaps most disturbing case, in 1957 the atmospheric observatory on Mauna Loa in Hawaii, under the direction of Charles David Keeling, began a series of very precise measurements of the concentration of CO_2 in the atmosphere. At that time, the concentration was 318 parts per million (ppm) per unit volume. Today, however, the concentration exceeds

350 ppm! This finding has been confirmed by measurements at numerous other sites around the world. It is indisputably true that in the last thirty years, there has been a significant and fairly regular increase in the proportion of CO_2 in the atmosphere. Older but admittedly more imprecise measurements suggest that the CO_2 concentration at the beginning of this century was 290 ppm. According to certain estimates, it might have been only 200 ppm at the maximum extent of the last glaciation.

This question of CO_2, however novel it might appear to public opinion, was in fact raised even before the beginning of this century by the Swedish scientist Svante Arrhenius. It is a relatively simple one: since people have begun burning coal and petroleum, we have introduced into the Earth + oceans + atmosphere system a new process of converting carbon into CO_2, in addition to the preexisting natural processes which also continue to operate. Today, about a dozen billions of tons of additional CO_2 are added to the atmosphere each year—and this represents only about half the amount produced by the combustion of fossil fuels, since it is believed that the other half must be extracted from the atmosphere by the oceans, by means of partially biological natural processes which are not yet fully understood.

CO_2 is used in photosynthesis, and one might assume that some of the additional CO_2 would tend to enrich the biosphere and increase the biomass. But deforestation, which also must have begun some 8,000 years ago, produces the opposite effect: since much of the wood that is cut down is burned, the result is, once again, to convert carbon into CO_2. The quantity is difficult to estimate since more is involved than just the vegetation being cut and burned: soils themselves can also become an important source of CO_2 when they are partly or completely denuded. The scale of this phenomenon is therefore considerable. It is estimated that the Earth is losing more than 150,000 km^2 [58,000 mi^2] of tropical forest every year. Today, the case of the Amazon forest is the most spectacular one (more than 120,000 km^2 [46,000 mi^2] lost in 1988), but we must not forget that in West Africa, more than half the forest—and perhaps as much as 80%—has been destroyed in less than sixty years, and that clearing operations still continue in Southeast Asia and Indonesia. The deforestation of China occurred a very long time ago; in America, vast stretches were deforested in the last century.

This destruction of the forests has major consequences. But in order to assess the gravity of the phenomenon correctly, we must not be misled by the prevailing alarmist theory that oxygen will disappear as the forests disappear. This idea betrays a misunderstanding of the way in which forests act as the "lungs of the planet." It is true that plants produce oxygen during photosynthesis, which is the basis of their growth; but they also consume it, by respiration and when they decompose. It is only because some of the dead vegetable matter ends up underground (where it is protected from oxidation) that there has been a slight excess of oxygen production and accumulation in our atmosphere, making animal life possible. Undoubtedly the disappearance of plant life would eventually lead to the disappearance of oxygen from our atmosphere, but that would take a very long time, and we would run into other serious problems well before it happened. The effects of rising CO_2 levels are perceptible in the absence of any detectable effect on oxygen, which constitutes almost 20% of our atmosphere. Note also that if all the wood that was cut down were used to build houses or furniture, or were simply buried, this would have no direct effect either on CO_2 or on atmospheric oxygen; and, moreover, if the trees involved were replanted, CO_2 would then be removed from the atmosphere. But in reality, almost all the wood is burned when land is cleared, and few trees are replanted.

The forest also plays an important role in the hydrologic cycle, since it restores to the atmosphere, by evapotranspiration, much of the water used by the plant; otherwise that water would run off and return almost directly to the sea. The forest thus helps maintain atmospheric humidity, and allows precipitation over regions far from the oceans. In Brazil, Eneas Salati has demonstrated by isotopic analysis that as one penetrates farther into the continent in Amazonia, the proportion of rainfall "recycled" in this way increases. Would the consequences of massive deforestation include relative desiccation of the Amazonian interior? Are such phenomena already occurring in Africa? A great deal of research is needed to answer these questions.

We can, however, accept as fact that the observed increase in the atmospheric concentration of CO_2 since the beginning of this century is a consequence of deforestation and the combustion of fossil fuels. The slight annual oscillation observed at stations in the Northern Hemisphere is easy to understand in conjunction with the growing season of Northern Hemisphere forests, which remove CO_2 from the atmosphere in summer and return it in winter. The fact that half the CO_2 released into the atmosphere does not remain there is still not completely understood on a quantitative level, but it is not really a mystery. As long as we continue to burn fossil fuels and fell trees without reforesting entire continents, this increase will continue. If we accept some very uncertain predictions about future use of fossil fuels, and if we assume that half the CO_2 produced by human activity will remain in the atmosphere, we find that a hundred years from now, the concentration of CO_2 in the atmosphere could double. Such an increase will certainly have a significant effect on climate, since CO_2 participates in energy exchanges between the Earth and space by means of what has been called the "greenhouse effect."

These well-established observations and conclusions are now being joined by new reasons for anxiety. It has been found that the atmospheric concentrations of other gases—so–called "trace gases"—have been rising much more quickly than in the case of CO_2, and that, despite their extreme rarity, these gases might contribute significantly to an intensification of the "greenhouse effect" and therefore to global warming. This is particularly true of methane (CH_4). Analysis of air trapped in Antarctic ice has demonstrated that the concentration of methane in the air has more than doubled over the past 300 years; it is believed that the increase is currently 1% per year, and that the concentration will soon reach 2 ppm. This is serious, since ounce for ounce, methane is a much more effective heat barrier than CO_2. Methane is released naturally in coal mines (in which, under the name of "firedamp," it occasionally causes disastrous explosions); some leakage also occurs around gas wells. But there are other sources in the biosphere: methane is produced in the stomachs of ruminants, and one of its major sources is cow dung. It is also emitted by rice paddies and marshes, in which it may spontaneously ignite to form "will-o'-the-wisps." It is also produced when plant material is incompletely combusted. Termites who chew up burned wood and stumps also excrete considerable quantities of methane. To what should we then attribute the increase in atmospheric methane? Extraction of fossil fuels? Increased cultivation of rice? More cattle-raising? Undoubtedly each of these factors plays its part.

Other gases can also contribute effectively to the greenhouse effect despite their extreme rarity in the atmosphere, especially nitrogen oxides and chlorofluorocarbons (CFCs). The latter have the same molecular structure as methane—hence the alternative name of chlorofluoromethanes (CFMs)—but the four hydrogen atoms are replaced by atoms of chlorine and fluorine. In the atmosphere, CFCs are exclusively of industrial origin. These products, manufactured by the American multinational corporation Du Pont under the Freon trademark, have been more and more widely and intensively used, in particular for refrigeration and air conditioning, for cleaning electronic components, and as aerosol can propellants. Their concentrations have increased from zero to a few hundred ppb (parts per billion), or a thousand times scarcer than

CO_2; nevertheless, their additional greenhouse effect might eventually be comparable to that of the added CO_2. They also appear to play a central role in the ozone question, which we will address later.

II

THE BALANCE

OF NATURE

AT RISK

Compared to the geological and astronomical rhythms that seem to have governed the great climatic changes of the past, the pace at which the global atmosphere is changing today is therefore extremely rapid. We can regard this change as a sort of "shock"—to use a term coined by the Russian atmospheric physicist Kyril Yakovlevich Kondratyev—that the climatic system is undergoing. How stable will the climatic environment be when faced with this very sudden perturbation of the "greenhouse effect"?

If we expand our perspective to take in the entire history of our planet, what is most striking is the relative stability of its climate. Of course it has undergone some major changes, especially those associated with continental drift and with the alternation of Ice Ages and warmer periods; and the Earth has also seen some true catastrophes, such as the one that ended the reign of the dinosaurs. But the biosphere did after all survive that catastrophe, otherwise we would not be here to wonder about it. We have evidence of the existence, for at least 3 billion years, of large quantities of liquid water (the oceans) on our planet. According to the prevailing theory of the internal structure and evolution of stars, which very convincingly explains the variety of stars we observe, the Sun's luminosity must have increased appreciably (perhaps by 40%) since that time, as a result of conversion into helium of part of the hydrogen in its core, and certain accompanying structural adjustments. With such a faint Sun, how was the Earth able to keep its water in liquid form? And if the Earth had ever been completely covered with ice, reflecting most of the incident solar radiation, would it not then have stayed that way forever, even if solar luminosity gradually increased? The key to this paradox (called the "faint early Sun" problem) seems to be precisely the "greenhouse effect," which depends on the composition of our planet's atmosphere, which in turn must have evolved over the eons. This brings us back to the problem of CO_2.

THE GREENHOUSE EFFECT

As this expression makes its way into everyday conversation, its exact meaning is being blurred. It does, however, describe a well-defined phenomenon which cannot be understood without some knowledge of the physics of atmospheric systems. And no matter how vivid the image might be, we must never lose sight of the fact that it is simply an analogy between a familiar phenomenon based on a well-established farming and gardening practice and a much more complex and poorly understood phenomenon which is the subject of fairly recent investigations by climatologists and planetologists.

We all know that a greenhouse is a device intended to protect or accelerate, or even to "force," the development of certain plants. Let us begin with an idealized, simplified—and in fact inaccurate—explanation of how it works:

The glass windows admit solar radiation, which is absorbed to some degree by everything inside the greenhouse, including the plants, of course. Like any object that is heated, the plants emit infrared radiation, which carries energy away and thus limits the extent to which they are heated. But the windows, which are transparent to visible solar radiation, are much less transparent to thermal infrared energy. The infrared radiation which is thus "trapped" inside the greenhouse helps maintain an elevated temperature before it finally escapes to the outside world.

The Earth's atmosphere, like the windows of a greenhouse, is transparent to visible radiation. Leaving aside a few watts per square meter absorbed in the stratosphere, most solar radiation can pass unhindered through the atmosphere, which consists primarily of nitrogen and oxygen in their molecular forms (N_2 and O_2). This radiation can then be absorbed by plants, the ground, the oceans, etc.; this is how the Earth's surface begins to be heated, and to radiate in the infrared. But the atmosphere blocks a portion of the infrared radiation, and thus, like the windows of a greenhouse, traps some of the absorbed heat. If we stick close to this simple mechanism, the analogy appears to be perfectly justified. But in actuality this phenomenon is obviously more complex.

We must not forget that in a gardener's greenhouse, the windows play an additional role, which most of the time is much more significant: they prevent the air that has thus been heated (and humidified by evapotranspiration of the plants) from leaving the greenhouse and being replaced by cold air from outside. In technical terms, there are no convective movements, or air currents, inside the greenhouse.

It is easy to observe this when a window occasionally gets broken: the convective movements start up again, and the temperature drops.

What about the Earth's atmosphere? The surface of the Earth can indeed lose heat as a result of multiple air currents, by means of turbulent convective movements with rising warm movements and descending cold ones (flows of "sensible" heat). It can also lose energy in the form of "latent" heat, which corresponds to the energy needed to make water evaporate. This heat is removed from the surface, where water evaporates, and returned to the atmosphere, where water vapor condenses. All of these movements in the atmosphere, with their varying degrees of regularity and violence, play an important part in the distribution of temperature within the atmosphere, and in its equilibrium. Now, although a convective phenomenon could exist entirely within a well-sealed greenhouse, it would in this case be too weak to have any substantial effect on temperature distribution—except in the case of a gigantic greenhouse, or a stadium covered by a transparent roof (it might be interesting in this connection to investigate the Houston Astrodome!).

How good, then, is the analogy between a greenhouse and the terrestrial atmosphere? To understand it, we must remember some basic facts about the vertical structure of that atmosphere. It is heated mostly from the bottom, where solar radiation is absorbed. In addition, its lower layer, which extends from the surface to an altitude of about ten kilometers, is the scene of continual mixing of masses of air, associated not only with unequal heating, but also with the Archimedean buoyancy forces which produce convection.

This region is therefore called the troposphere, from the Greek *tropos* meaning a rotation. In this layer, temperature decreases with altitude (by about 5–10°C per kilometer [3-6°F per thousand feet]) up to a level called the *tropopause*: literally, the level at which rotation stops. The altitude of this level depends—very approximately—on the intensity with which the tropospheric "pot" is heated, and thus on the latitude and the season. At the equator, the tropopause lies at 16 or 17 kilometers [about 10 miles], and its temperature is very low (down to –90°C [–130°F]); at the poles, by contrast, it is seldom higher than 8 kilometers. Similarly, at our own temperate latitudes it is higher in summer and lower in winter. Above it lies the stratosphere, in which the temperature, which is initially constant, increases with altitude up to a maximum of about 0°C [32°F] at the stratopause, about fifty kilometers [thirty miles] above the ground. These layers are heated from above, by absorption of ultraviolet radiation. This stratification, with a temperature which increases with altitude, is relatively stable: relatively little vertical motion occurs within it.

Now that we have established the limits of the analogy, we must add some more definitions in order not only to understand the difficulties faced by scientists, but also to appreciate the real extent of the danger that politicians must take into consideration when making their decisions. We know that the Earth's atmosphere consists of gases: essentially nitrogen (78%) and oxygen (21%). These two gases, consisting of the symmetrical diatomic molecules N_2 and O_2, do not interfere with infrared radiation. All the polyatomic molecules, however, such as carbon dioxide (CO_2), water (H_2O), and ozone (O_3), have vibration and rotation modes which make them efficient absorbers of this infrared energy. If we examine the spectrum of the radiation emitted by the Earth back towards space, we find that a portion of it is blocked by these molecules. And if we look upward from the Earth's surface, we find in this part of the spectrum a significant *downward* radiation, emitted by these atmospheric molecules; this means that astronomers cannot observe the Universe from the ground at these wavelengths. The climatic equilibrium that we are familiar with therefore depends first of all on the "greenhouse effect" resulting from the water vapor and CO_2 which are vital constituents of the natural atmosphere. The temperature at the Earth's surface can thus be maintained at a comfortable level—approximately 15°C [59°F]—precisely because the 390 W/m² emitted by the surface back into space is partly compensated for by approximately 300 W/m² radiated back downward by the atmosphere in the infrared.

We have also observed the existence of a "window," specifically at wavelengths between 8 and 13 micrometers, where the radiation is not blocked. Through this window, astronomers can study objects which they refer to as "cold" (less than 1500 K), and which radiate in the infrared. It is also through this window that most of the energy radiated by the Earth passes, giving rise to a certain degree of equilibrium. When we add a little more CO_2 to the atmosphere, the window gets a little smaller. When certain minor constituents are added—such as methane, CFCs, nitrogen oxides—the effect is like putting up shutters where there were none before. The changes in the composition of the atmosphere that are presently occurring must therefore inevitably enhance the greenhouse effect, in other words intensify the trapping of energy and raise the temperature in the lower atmospheric layers and at the surface. This intensification of the greenhouse effect will necessarily result in global warming. Clearly, we are now living in a transitional phase leading towards a new climate, with higher temperatures at the Earth's surface.

Is this "greenhouse effect," as sketchily described above, unique to the Earth? The answer, as we have known for some time, is No: this effect is observed in every planetary atmosphere, all of which contain polyatomic constituents. The phenomenon is particularly pronounced in the atmosphere of Venus, which is very rich in CO_2; the planet's surface temperature is several hundred degrees, an inferno that rules out any form of life. Soviet probes have simply confirmed what was assumed for years. However, two factors peculiar to the Earth's atmosphere must be taken into account if we are to understand the complexity of the phenomenon. The first concerns the existence on our "blue planet" of a considerable oceanic mass which exchanges heat and matter with the atmosphere, especially in the form of water vapor, so that temperature changes necessarily lead to modifications in the hydrologic cycle. And these changes turn out, in many ways, to be more important than the temperature changes that can be attributed directly to enrichment of the atmosphere in CO_2 (and methane, CFCs, etc.). If we start to raise the surface temperature, this intensifies evaporation and thus increases the quantity of water vapor in the atmosphere. But this water vapor absorbs over a large portion of the infrared spectrum, especially in the wavelength ranges of 5–8 micrometers and 16–50 micrometers, while CO_2 absorbs principally in a narrow band around 15 micrometers. The result is a further intensification of the greenhouse effect due to additional humidification of the atmosphere. In calculations which predict between 2 and 4°C of warming as a result of a doubling of atmospheric CO_2, a large part of this warming can in fact be attributed to this "water vapor feedback."

The second factor unique to the Earth's atmosphere— and perhaps an even more decisive one—concerns the existence of clouds. Of course there is no lack of clouds on other planets, but it is only on Earth that they play a role that is both important (much more so than on Mars, where they are rare) and variable, since they do not cover the entire planet (as they do on Venus) but only between a half and two-thirds of its surface. The reason is that clouds are very good reflectors of solar energy. It is therefore usually believed that if cloud cover is increased, the Earth's surface will then be cooled. That is often in fact true, but at the same time, clouds also block infrared radiation. We know that a cloudless night can be extremely cold, while a cloudy one is quite mild. In our own temperate regions, this is the essential role of clouds in the radiative balance: in the polar night, it is their exclusive function. Here, then, is a vital factor defining climates and their evolution, which upsets the simple picture of a cloud which blocks the Sun and cools the Earth. The study of cloud cover is only in its infancy, however, especially when it comes to understanding its role on a global scale. The formation and dissipation of clouds, and their influences on radiation interchanges, involve extremely complex phenomena. The reflectivity (or *albedo*) of clouds depends on their thickness, the number and size of the water droplets or ice crystals that they consist of, and possibly the various kinds of pollutants that they contain. Their contribution to the greenhouse effect, as far as the planet as a whole is concerned, depends primarily—in addition to these parameters—on the altitude of their tops.

High clouds such as cirrus and cumulonimbus, with very cold tops close to the tropopause, radiate very little in the infrared; they therefore have a very powerful greenhouse effect, since they prevent the infrared radiation emitted by warmer surfaces from escaping into space. In cumulonimbus clouds, which reflect sunlight very well, the two effects approximately balance each other out. With cirrus clouds, which are often very thin, it is generally the greenhouse effect (that is, the warming effect), which predominates. Low clouds, on the other hand, emit almost as much infrared as the surfaces below them, which means they contribute little to the greenhouse effect. These clouds—especially the extended decks of stratus and stratocumulus found above the oceans—essentially play a cooling role, since they reflect most solar radiation and thereby prevent the ocean's surface from absorbing it.

It is a very tricky matter to assess correctly the net effect on today's climate of these two contradictory influences produced by cloud cover. If the climate begins to change as the Earth's surface becomes warmer, what changes will occur in cloud cover? How do we find out whether these modifications will reinforce or attenuate the initial trend? We must determine, among other things, whether and where there will be more or fewer low or high clouds. Only recently have we acquired some elements of an answer to these questions, thanks to the global view now offered to us by satellites, and the numerical simulations made possible by supercomputers.

THE OZONE QUESTION

The phenomenon referred to as the "destruction of the ozone layer," now largely hijacked by the media, can only be understood in the context of this system of complex multiple interactions. This phenomenon must be put back in its proper context, since I have perceived a tendency to exaggerate its intrinsic importance. We know that the ozone layer, which like the stratosphere is located at altitudes of between 15 and 50 kilometers [10 and 30 miles], is created by the action of solar radiation on oxygen molecules in the atmosphere. More precisely, the relevant spectral region is ultraviolet, a type of radiation which does not transport much total energy, but in which each individual photon is energetic and can interact with molecular structures. This interaction first of all involves absorption of some of the ultraviolet radiation, which heats the stratosphere from above: it is precisely because this layer of the atmosphere is heated from above that it is stable and stratified, since buoyancy forces cannot act effectively within it (hence the name "stratosphere").

Absorption of these ultraviolet photons in the stratosphere is accompanied by dissociation of a small fraction of the oxygen molecules (O_2). The result is that a certain number of free oxygen atoms can then associate with other O_2 molecules to form molecules of ozone (O_3). On the whole, the ozone layer can be regarded as a thin film which has the ability to absorb solar photons in the spectral region called the "near ultraviolet," namely wavelengths just beyond the visible. Without this layer, these photons would reach the Earth's surface, and the problem is that they are sufficiently energetic to disrupt the structure of complex, fragile molecules, especially the molecules of living organisms. In fact, it is believed that life could not have emerged from the sea until an ozone layer had formed. We can therefore see the considerable hazard that would exist for life on Earth if the ozone layer were destroyed. And it is feared that any diminution in this layer would produce disturbances—whose exact nature is still very difficult to predict—in ecological balances, along with a significant increase in the incidence of cataracts and skin cancers, both of which are associated with exposure to near-ultraviolet radiation. These fears, and not the fact that ozone is also a gas which contributes to the greenhouse effect and therefore plays a role in climate, are what has captured the attention of politicians and the media.

Stratospheric ozone protects life, but ozone itself is an irritating gas that is dangerous to breathe and is equally harmful to plant life. Aircraft flying in the stratosphere—especially on north polar routes in the months of May and June, when ozone reaches a maximum—must carry equipment designed to limit ozone concentrations in the cabin air to low levels, for the protection and comfort of both crew and passengers. In the troposphere ozone is a pollutant, a component of the smog that is formed by complex photochemical reactions (processes which involve solar radiation photons). Increasing levels of tropospheric ozone definitely should be a concern, at least in the industrial regions of the world, since they pose a serious public health problem. If we compare observations made a century ago at the Montsouris Park Observatory in Paris with current measurements, we find that the quantity of ozone has almost doubled.

But the "ozone question" still refers to *stratospheric* ozone and its possible destruction. Concern began to grow in the late 1960s, with the prospect of a considerable increase in commercial air traffic in the stratosphere. In addition to the water vapor and CO_2 that result from any hydrocarbon combustion process, jet engines produce a certain volume of nitrogen oxides, which can cause the breakdown of ozone—or at least that was the theory in about 1974. In any event, the results of the first investigations led to cancellation of an American project to develop a supersonic transport, and to suspicions about the Concorde. In reality, it was later concluded that at Concorde's flight altitudes the effect might be the opposite: a slight increase in ozone! Stratospheric chemical reactions are extraordinarily complex and, even today after more than fifteen years of intensive research, we do not know enough about the constants which govern the rates of these reactions and thus determine how they interact.

Another threat to the ozone began to emerge in the 1970s, when two researchers in California, Sherwood

Rowland and Mario Molina, brought to light the role of CFCs. These compounds, also called "Freons," are justifiably valued for many applications because of their enormous stability. In the troposphere almost nothing can destroy these molecules which are produced exclusively by human activity. As human beings have used aerosol cans and discarded old refrigerators over the years, these gases have migrated up to the stratosphere, and they continue to do so. Ultraviolet photons in the stratosphere are capable of "breaking" the CFC molecules, which yields chlorine oxide and free chlorine atoms; it has been found that these chlorine compounds are very effective at destroying ozone. So here we have one more complication in the physics and chemistry of the stratosphere. Despite all the uncertainties, the threat is a serious one.

Scientists therefore expected to observe a global, progressive decrease in stratospheric ozone, a phenomenon that would be extremely difficult to detect since the concentration of this gas is highly variable as a function of latitude, season, meteorological structures, and even solar activity. This progressive decrease is starting to be observed, and it does seem to be attributable largely to CFCs. The big surprise was an announcement in 1985, by a team headed by Joe Farman working at the British research base at Halley Bay in Antarctica, of a radical decrease in the ozone layer above the south polar region during the Southern spring. These researchers found that this "ozone hole" had in fact been forming for several years, expanding each spring and then filling in when the huge "polar vortex" breaks up at the start of the austral summer. This discovery aroused a great deal of interest, and the international scientific community in this field, already highly motivated by the ozone question, immediately attacked the problem: since 1986, expeditions have been organized to make more complete *in situ* measurements in Arctic and Antarctic polar regions. The telltale presence of chlorine oxides, originating from chlorofluorocarbons, has been confirmed. In the South, the trend is rather disturbing: the hole has reappeared with greater intensity each spring, and there are even some indications that the ozone-deficient area is getting larger. In the North the phenomena are not so dramatic, but there is also a steep drop in ozone close to the pole.

We are beginning to understand what is going on, or at least we are beginning to think we do. But the measurements, analysis, and discussions continue. During the polar winter, the air circulation in the stratosphere above the pole organizes itself into an enormous eddy called a "vortex," which effectively isolates this region from the rest of the atmosphere. In the absence of any solar radiation, and with a very low level of infrared radiation from below, the temperature drops considerably; this allows the formation of clouds, which are normally very uncommon in the extremely dry stratospheric air. The vortex and the clouds persist for part of the spring, and certain highly complex chemical and photochemical reactions occur on the particles making up these clouds. The chlorine compounds produced by CFCs appear to participate very effectively in these reactions, destroying almost half the ozone in the vortex. As summer approaches and solar radiation increases again, stratospheric temperatures rise, the clouds disappear, the vortex breaks up, and the air mixes: this restores the ozone. The formation of the ozone hole thus appears to be closely associated with the development of the polar vortex, and it is difficult to imagine how the vortex might become more extensive than it is now. As a result, there is no threat that all the ozone in the world's stratosphere might disappear into the dreaded hole. Nevertheless—confronted with a disturbance which is indisputably of human origin, Nature did not react as predicted. And although we are now achieving a better understanding of what we had forgotten to take into consideration, can we be so sure that Nature does not have some other surprises in store for us?

IS THE THE WORLD GETTING WARMER?

Will all these phenomena—higher levels of atmospheric CO_2 and methane, the introduction of CFCs, and more—result in a climatic New Deal? Climate is a fundamental aspect of our human environment, and determines on a broad scale the opportunities for life on our planet. It would be a very good idea to try and understand the processes that are occurring, so we can see where our own actions are leading us. No matter how breathtaking the models currently being created and popularized as a foundation for future scenarios, the remaining uncertainties must not be concealed. On the contrary, it is only by taking these uncertainties into account that we will be able to evaluate the often emotional debate that is unfolding on the public stage with regard to the dangers threatening the Earth. We will see that in some cases, this debate is not without elements of hypocrisy and political calculation.

Nostalgia for a Golden Age and apprehension about the end of the world—both of them perhaps a matter of climate—have existed for a very long time. Is the biblical Flood a remembrance of a relatively rapid rise in sea level 6,000 years ago, exceeding today's level by 13 feet and submerging the Nile Delta? Jacques Labeyrie presents this hypothesis in his book *L'Homme et le Climat*. Farmers in the Savoie region of France who had to abandon their villages to advancing glaciers during the "Little Ice Age" of the 17th century knew nothing of the great glaciations of the past, and certainly must have wondered what was happening to them. In fact, the question of climatic change has been a matter of scientific inquiry for less than two centuries, in other words since we acquired a better understanding of our planet's history and the physical mechanisms which govern its climate.

Once the traces of past great glaciations had been recognized, the question of Earth's future consisted for many years of wondering if and when the ice would return. Today the scientific community is leaning towards the ideas developed by Milanković, favoring an astronomical origin for the multi-millennial rhythms of the Ice Ages (based on quasi-periodic variations in the Earth's orbital and rotational parameters)—although it is not ignoring the role of other mechanisms such as the greenhouse effect. Although there is a great deal more to learn about this subject, it is reasonable to assume that the glaciers will return in a few millennia or tens of millennia.

Smaller and more rapid climatic fluctuations have been observed, especially the "Little Ice Age" which began in the 17th century and, according to some researchers, lasted until 1850. There are some indicators suggesting a link between these fluctuations and variations in solar activity. This link, which is still hypothetical, is difficult to understand, since the Sun's activity—taking into account all the phenomena associated with the sunspot cycle, the solar and interplanetary magnetic field, solar flares, and the solar wind—does not seem to have a perceptible and lasting effect on its luminosity. It is true that precise measurements of this luminosity, obtained from satellites outside the atmosphere, have been available for only about twenty years. But have there in fact been significant variations (on the order of 0.5% or more) in the Sun's luminosity on a scale of centuries, millennia, or millions of years? Theoreticians specializing in solar physics have no unequivocal answer to this question. The types of radiation which vary cyclically (in 11-year and 22-year cycles) and abruptly (on a scale from several hours to several days) with solar activity—such as X-ray and ultraviolet photon emissions, radio waves, and energetic particles—have many significant effects on temperature and ionization in the upper and middle layers of Earth's atmosphere, but they do not, on the whole, transport very much energy, and it has not by any means been demonstrated that their effects penetrate into the troposphere, the real stage on which meteorological and climatic events are played out.

The search for correlations between solar activity and the phenomena of meteorology and climate continues to attract many scientists. It is worth noting, however, that none of the correlations announced in the past has stood up to critical examination over a long series of measurements (more than 40 years). This does not always prevent the "heliocyclophiles" from trotting out old series of data that have been discredited for years, like the claimed correlation between the solar activity cycle and the level of Lake Victoria, which did in fact match up between 1880 and 1928 but have been uncorrelated ever since. Personally, I believe that the energy expended in chasing after these chimeras would be better applied to investigations of the internal variability of the Earth + atmosphere system. It is true that the Sun is essentially the only source of the energy that makes this system run, but everything indicates that the solar energy flux is hardly affected at all by the epiphenomena associated with our star's activity, and that explanations for meteorological and climatic variations should be primarily sought on Earth. The "Earth system"—which is much more complex than the "Sun system"—contains far more mechanisms for variability on different time scales.

Only thirty years ago, scientists were still worried about the return of the glaciers, even imagining a very rapid return in the form of what was called a "snowblitz": a massive winter snowfall over enormous land areas, reflecting solar radiation and leading to persistent cold with greater and greater accumulation of snow cover. The average temperature at the Earth's surface had in fact started to decrease during the decade 1950–1960, after a period of irregular increase between 1850 and 1945. Researchers were also worried about the effects of an accumulation of aerosols in the atmosphere, essentially droplets of sulfuric acid (the same ones responsible for acid rain), formed from the sulfur dioxide (SO_2) produced by combustion of sulfur-laden coal and petroleum. These aerosols might increase the proportion of solar radiation reflected to space, at least above ocean surfaces, and thus reduce the amount available to heat the planet. This "human volcano" seemed to rival in magnitude the natural volcanic activity that intermittently generates temporary hazes of this kind, which do indeed cool the Earth.

Today, the hypothesis that is most commonly accepted—and feared—involves global warming of the Earth's surface, associated with an intensification of the greenhouse effect. This scenario is in fact a composite structure built up from three stages of modeling. The first is an economic and industrial model, which predicts future rates at which CO_2 and CFCs are injected into the atmosphere; the second is a biogeochemical model that predicts how the atmospheric concentration of these gases will develop, taking into account predicted injection rates and the processes by which these gases (and others such as methane and nitrogen oxides) are exchanged among the atmosphere, the oceans, the soil, and the biosphere; and lastly there is a climatic model, used to predict how the climate, with its atmospheric and oceanic components, will change in response to these modifications in atmospheric composition. Using certain projections of economic activity and certain models of biogeochemical exchanges, it has been concluded that the effective concentration of greenhouse gases in the atmosphere might double by the year

2030. Some climate models then predict an increase in average temperature at the Earth's surface of between 2 and 4°C [4 and 7°F].

This warming, which might seem modest, would have substantial consequences, especially if it took place within less than a century (which is a very brief adjustment period for ecosystems that are well adapted to the present climate). A mean global temperature increase of 4°C [7°F] is in fact quite comparable to what has happened since the last glacial maximum. It would raise the average temperature of the Earth's surface to 19°C [66°F], which is unprecedented in the planet's recent history. We need to go back several million years, perhaps as far as the Miocene, to find the same value.

What would be the consequences of such a change? The warming would not be equally distributed over the planet, nor over the whole year. Most models predict that the climate would be more sensitive at higher latitudes, with temperature increases as high as 6–10°C [11–18°F] near the poles. This prediction is based on two mechanisms that are fairly well represented in the models. The first concerns the greater stability of the atmosphere near the poles: an intensification of the greenhouse effect could produce warming of the lower layers, which would not be limited by more intense convection. The second mechanism is known as "ice–albedo feedback": if the surface starts to heat up, the snow or ice covering it will tend to melt. This then has considerable repercussions on energy exchanges between the surface and the atmosphere. Snow and ice have a high albedo, meaning that they reflect much of the solar radiation that strikes them. If warming caused some of this layer to disappear, exposing a darker surface, the quantity of solar energy absorbed would rise. The phenomenon might be particularly pronounced if sea ice melted, since water has a very low albedo. Increased absorption would cause additional warming, so that ice would form more and more slowly during the winter. We would then see an amplification in the climatic change, a process called "positive feedback."

How extensive would these changes be? If, as some researchers believe, the entire ice cap covering the Arctic Ocean disappeared, Archimedes' principle tells us that conversion of this floating ice into water would not affect sea level; but it would completely upset the regional and global climatic situation. In addition to changes in the absorption of solar radiation, we might see a considerable increase in oceanic evaporation. And with more moisture in the air, would there not be more cloud cover, blocking the Sun and partly or completely counteracting the effect of the decreased surface albedo? Or would the greenhouse effect of those clouds predominate, leading to enhanced warming of the Arctic? Would there be more snowfall on continental areas around the Arctic Ocean? And would this not be a factor leading to a recurrence of the Ice Age? It is difficult to say, since the representations (or "parameterizations") of these various processes in our climatic models are for the most part very rough approximations, and therefore not very reliable.

Other spectacular consequences are often mentioned: if the continental ice caps melted, the result would be a rise in sea level that might submerge all the coastal plains. It is true that many mountain glaciers have been retreating since 1850, which might explain a 3–4 centimeter rise [1.5–2 inches] in sea level since that date, but it is not enough to explain the 9–10 centimeter [3.5–4 inches] value that has apparently been observed. Current worries are focused more on the Antarctic ice cap, especially on its "western" portion (south of Tierra del Fuego) which, according to some researchers is relatively unstable. If these two million cubic kilometers of ice were to melt or slide into the ocean, the result would be a 6-meter [20-feet] rise in sea level! That would certainly be enough to submerge some densely populated regions such as Bangladesh, which already suffers terribly from each major cyclone (hundreds of thousands died in 1970). The Dutch already have an infrastructure of dikes to protect their coastline and would not find it too difficult to increase their height. But there would be problems over much of the coastline of Europe and the world. Would Venice, already seriously threatened by pollution and flooding, finally be engulfed? Perhaps New York would be transformed into a new tourist attraction: Broadway and Fifth Avenue would make magnificent canals, and Manhattan island would be split into several islets. And let us not even contemplate what might happen if *all* the continental ice—32 million cubic kilometers of it—were to melt...

This is all very spectacular, but it does not take into account the time scale of these changes. At the very least, it is important to know if these scenarios will unfold within a century or a millennium. It is true that the atmosphere can adjust rapidly to any perturbation in energy transport conditions and therefore to an intensified greenhouse effect. But it does take a certain amount of time to make ice move. At the end of the last glaciation, sea level rose by more than 100 meters [330 feet], but it took about 10,000 years to do so. It is therefore not at all inconceivable that sea level might rise 6 meters [20 feet] in two or three centuries. However, a great deal can happen in two or three hundred years: coastlines can be protected or areas can be abandoned, depending on their importance. In New York, skyscrapers are demolished every few decades to give way to new buildings; it would not be difficult, it seems, to adapt to a rising sea level. Is a faster rise possible, perhaps resulting from a catastrophic slippage of two million cubic kilometers of ice [a block of ice 1,000 miles on a side and half a mile thick] from western Antarctica into the sea? Most specialists do not take this threat seriously and believe instead that the rise will be limited to about a meter by the end of the next century, although estimates range from 20 to 150 centimeters [8–60 inches]. Moreover, this rise would be mostly the result of the thermal expansion of seawater.

The specter of melting ice is an evocative one, but it would not necessarily be the most serious consequence of global warming. As I have said, there is no question that surface warming would produce a general redistribution of precipitation. The predictions of the models do not agree on this point, but undoubtedly certain dry regions would no longer be dry, while others would experience increased precipitation. The consequences—not just ecological, but also economic and human—might be very far-reaching indeed. For example, some researchers already think they can predict the "Mediterranean-ization" of France. This would obviously unleash a torrent of problems relating to stock-raising, hydroelectric power generation, and certain kinds of agriculture. But other activities, such as tourism, might benefit from it.

As in the case of the melting ice, the critical question here is the rapidity of the change. It takes a certain amount of time for ecological adjustments to occur, for one type of vegetation to replace another. In Europe and North America, the various vegetation zones have followed—sometimes not without difficulty—the irregular northward retreat of the ice sheets over 10,000 years of deglaciation. Some cold-adapted species became trapped, and now survive only on the north-facing slopes of hills and mountains. If the Earth became warmer, they would be doomed. If the climate changes too rapidly, can the biosphere keep up with it? Or would the climatic shock associated with an intensified greenhouse effect, for which humans are responsible, lead to the disappearance of a very large number of species? This is a source of profound anxiety.

MODELS

These audacious predictions, now bolstered by numerical modeling on mainframe computers, are still shaky; they may prove misleading if we do not very scrupulously take into account the different time scales governing the processes in question and the various confidence levels that can be attributed to models of the various processes. We have already mentioned the three types of models involved: economic development models, biogeochemical cycle models, and climatic process models. Most specialists in climate modeling prefer not to address the difficulties inherent in predicting changes in CO_2 concentration. They calculate the new equilibrium climate that should exist with an atmosphere having twice the current level of greenhouse gases (with reference to CO_2, although we must not forget methane, CFCs, etc.). This is called a "climate sensitivity" calculation, which must not be confused with a calculation of predicted change. The result is an average global warming which some estimate at 2°C [4°F], others at 5°C [9°F]. It is evident that even this "simple" calculation conceals uncertainties, particularly those associated, as we have seen, with feedback problems such as ice–albedo feedback and feedback from cloud cover to radiative balance due to the opposing influences of greenhouse effect and albedo effect.

To determine the date at which this warming will occur, we must know how many years it will take before the quantity of CO_2 in the atmosphere in fact reaches 600 ppm (approximately twice the concentration in the year 1900); then, once this concentration has been reached, how much time it will take for the climatic system to reach its new equilibrium.

How can we obtain a precise answer to the first question? Economic modeling, which we specialists in the exact sciences often tend to regard with scorn (and rightly so, sometimes), is in fact extremely difficult because it involves modeling a radical change in the system—doubling the world population in less than 100 years, doubling economic activity in less than 60 years. An error of 1% per year in the rate of economic growth can result in an error of 30 years in the date at which we reach 600 ppm CO_2; and we must also take into account choices among different energy production technologies, progress in the efficiency with which energy is used to produce consumer goods, etc., etc. The industrial CO_2 emission scenario is therefore highly uncertain.

Moving on to a later phase, what changes will occur in the fraction of this supplementary CO_2 that remains in the atmosphere, as the latter becomes enriched in CO_2? Those whose business it is to model biogeochemical cycles find it very difficult to understand in detail how it is that almost half the CO_2 added to the atmosphere since 1900 has ended up in the oceans. It appears difficult to account for this without assigning enormous significance to biological processes which apparently incorporate some of this carbon into fecal matter and send it directly to the ocean floor. If the temperature and chemistry of the ocean's surface layers begin to change, will these processes be affected, and how?

If, despite all these uncertainties, we accept as a working hypothesis a particular scenario involving enrichment of the atmosphere in CO_2, the modeler can then attempt to calculate how the climate will change during the next few decades. We must then bring into the picture a factor which is still enigmatic: the inertia of the system, essentially associated with the role of the oceans. Climate modelers must therefore turn their attention to this linkage between the atmosphere and the oceans, not forgetting the role of ice. An intensification of the greenhouse effect means that infrared radiation becomes trapped in the lower layers of the atmosphere, which heats up the surface. There is, of course, an oceanic surface layer which interacts directly with the atmosphere and which will therefore be warmed by those few additional watts per square meter. Since the volume of this surface layer is limited, its temperature can respond relatively rapidly, as it does in the course of the seasons. But such would not be the case at all if the entire volume of the oceans had to be warmed up. In the final analysis, however, it is indeed the entire volume of the oceans that must be warmed up, and that leads to a delay in global warming, a delay whose magnitude depends on the circulation of deep ocean waters.

The particulars of this circulation are still rather poorly understood. We know that it is very slow: complete mixing can take centuries, indeed more than 1,000 years in the Pacific. Our knowledge has advanced considerably in recent years as a result of atmospheric nuclear tests in the 1950s and 1960s. These tests created, almost instantaneously, a certain number of radioactive isotopes which did not exist previously, or existed only in very small quantities. These isotopes are being used as "tracers" of the oceanic circulation: it is easy to track their penetration into the ocean, to see where they go over the years, and hence to determine the circulation rate.

Although surface ocean currents depend essentially on winds, the "engines" that drive deep-ocean circulation are contrasts in temperature and salinity. When seawater freezes, the resulting ice contains relatively little salt. The formation of sea ice around Antarctica every winter thus leaves behind a volume of water that is very cold and salty, and therefore denser; this water flows downward and then forms a very distinct layer which spreads over the entire ocean floor, reaching as far as the Northern Hemisphere. Deep water also forms at the edge of the North Atlantic ice-pack, and then flows at depth towards the south, with a total current volume equivalent to twenty times the flow of every river on the Earth's surface. At the Straits of Gibraltar, Atlantic water enters at the surface; lower down, it is the much saltier Mediterranean water which pours out towards the bottom of the Atlantic. In the Atlantic, it appears that ocean currents as a whole transport heat from south to north; a portion of this heat comes (especially at the surface) from the Indian and Pacific Oceans, having flowed around Africa. This flow is balanced by the deep, cold current flowing south. What is more, these various deep-water and surface currents can carry not only heat—the key factor in the inertia of the climatic system—but also gases; this complicates the problem we already face in determining the scenario for CO_2 enrichment of the atmosphere.

At every stage of the modeling process we therefore encounter processes which we still understand very poorly, and which each constitute a source of uncertainty. In addition, these various stages interact with one another. Temperature changes necessarily have an effect on gas exchanges between the atmosphere and the oceans, and therefore on the evolution of CO_2 concentrations in the atmosphere. The effects of climatic changes on the biosphere will have feedback effects on CO_2 exchanges between the biosphere and the atmosphere, but also on the climatic processes themselves. If vegetation changes, this will modify evapotranspiration and the Earth's albedo. Certain types of phytoplankton emit particles (aerosols) which contribute to cloud formation in the atmosphere. If new conditions favor the development of these plankton, this biological process will affect the already complex feedbacks that govern cloud cover. Lastly, climatic changes will have an impact on economic activities and on energy production, if only in terms of heating requirements, which will in turn modify the rate at which CO_2 is produced. Modeling these impacts on the various activity sectors adds yet another level to the scaffolding of our models. And, of course, the financial support offered for much of this research is largely motivated by the desire to obtain predictions which, if they are considered sufficiently reliable and alarming, will dictate changes in our future economic policy and thus change rates of CO_2 production. Thus the final feedback effect is the involvement of human intelligence.

THE MYTH OF EQUILIBRIUM

The word "equilibrium" has found its way onto these pages several times. But although this concept is essential in physics, when applied to the environment it is an equivocal one that risks spreading considerable confusion about researchers' theoretical activities. It is the foundation of many strongly held ideological and political positions which unjustifiably invoke the authority of science. Is it not obvious from the examples that we have analyzed that the idea of equilibrium, of "balance," is always a relative one that depends on the time scale of the phenomenon in question? We have also shown that climatic change involves time scales that vary enormously depending on the process. We can safely say that if we look at the planet as a whole, we will always see phenomena occurring outside a state of equilibrium.

When we examine the "Earth system"—with its solid part, its liquid part, its living part and the atmosphere—we recognize clearly that each of its constituent elements has a time scale of its own. For example, the solid part changes very slowly: continental drift and mountain-building take tens and hundreds of millions of years. Nevertheless, this solid part reacts to what is happening above it. We know, for instance, that although the last Ice Age ended more than 10,000 years ago, the Earth's crust in Scandinavia, which was greatly compressed at the height of the glaciations because of the weight of ice on top of it, is still bouncing back: this means that on the Norwegian coast it appears that sea level is dropping! No doubt one could argue that this is a persistent trace of the former equilibrium, and that the current elevation of the crust following the disappearance of the glacial load is a transition towards a new equilibrium. But will the crust achieve this new equilibrium before the advent of another climatic change, with its new load of ice? Did an equilibrium really exist before? After all, the last glacial maximum followed a period in which there was no ice at all. In fact, every event represents a departure from equilibrium, if that condition is understood—as it has been for too long—as a static equilibrium.

As far as the ocean is concerned, we have already mentioned the differences between the rapid circulation of surface water and the slow circulation of abyssal water. This means that a portion of the Earth's oceanic water was in contact with the atmosphere at the time of the Little Ice Age, when temperatures were colder. And this period in turn followed a warmer period. The very fact that climatic changes have occurred, on a time scale shorter than the time necessary for ocean mixing, necessarily implies some imbalance among the atmosphere, the ocean surface, and the deep layers of the ocean. The ocean (and not the water itself) therefore possesses a thermal memory which would not exist if the system were at equilibrium.

Such is also the case at a time scale with which we are very familiar, namely the seasons. From an astronomical point of view, at our middle latitudes the day of the summer solstice is the day on which we in Paris, New York, or Moscow receive the strongest solar radiation, taking into account the length of the day and the Sun's height in the sky. All during what we call summer, the solar energy flux then decreases. However, the temperature maximum is generally reached in July, or even in August, when insolation is weaker than on June 21. The reason is precisely the fact that there is no equilibrium, no immediate adjustment to solar irradiation; instead, heat needs to be stored. This delay is most noticeable near the sea, specifically because volumes of water can mix, and the water must therefore be heated to a certain depth (several dozen feet). Once again, then, we almost always find a situation that is not in equilibrium, and the water at the beach is often warmer in October than in June. On the continent, however, and particularly in places far from the influence of the oceans, the temperature rises and falls during the day and during the year, with relatively little lag behind the Sun.

We can see, then, that even if the atmosphere alone "wanted" to come quickly to an equilibrium as the greenhouse effect intensified, it could not, since over two thirds of the Earth's surface it is in contact with the oceans, exchanging heat and moisture (and CO_2) with it, and thus being affected by its thermal inertia. Calculations indicate that these effects could delay warming by several decades, and even longer in the Southern than the Northern Hemisphere, because of differences in land/water surface areas. But these results are still weighed down with uncertainties.

Another equilibrium myth concerns the biosphere. Take the case of a forest: we all know that when a forest fire occurs, it takes a certain amount of time before the forest re-establishes itself; during that entire period, a situation of non-equilibrium exists. And this succession of non-equilibrium states can be entirely natural, in fact necessary to the stability of an equilibrium if the word is understood in its wider sense. In the past, before demographic pressure became too great, brushfires in tropical regions were part of a system of alternating cultivation: farmers would burn an area, cultivate it for a few years, then leave it fallow for ten years or so while the tropical forest reclaimed its rightful place. But if the land is pushed too hard and not allowed to rest, the fragile tropical soils often end up becoming completely sterile. At other latitudes the soils may prove to be less delicate. It is true that a forest takes a much longer time to re-establish itself, but natural fires clean out some of the vegetation and make the system as a whole more robust. A succession of small disasters can make the real catastrophes less common. The fires which devastated Yellowstone National Park in the summer of 1988 took on catastrophic proportions not only because of the extreme heat and dryness of that particular summer, but also because, for many years, the park had been run following a policy of suppressing every fire, even those of natural origin. The result was a huge accumulation of dead wood and other flammable material.

A number of species have tried to take advantage of the new climatic situation that has prevailed since the end of the last glaciation, including *Homo sapiens*, who invented agriculture and started cutting down forests and irrigating deserts, causing certain specific imbalances. In some regions of Asia, it is impossible to say what the "natural" vegetation was before agriculture. It would be an illusion to believe that equilibrium was suddenly disrupted; in reality, as I have just mentioned, even before the invention of agriculture there was already no equilibrium. So although the idea of the "balance of nature" has been conceived and has been a part of scientific and philosophical thinking for years, it is because in the course of a generation there did not seem to be many changes in the relationships between humans and their environment, and in the environment itself. Today things are very different, and a new concept is needed. It can no longer be claimed that we can exploit Nature without profoundly upsetting it on a global scale. But that is no reason to think that, without human beings, a "balance of nature" would prevail. Today this idea of equilibrium is obviously a myth, if we view it as a static equilibrium. In this world, all is change, and change is not inherently a bad thing. We must therefore come around to the idea of a *dynamic* equilibrium which can take into account the various developments that affect our planet, in terms of its atmosphere, oceans, ice cover, biosphere, and human activities.

But that is not all. We must also take a closer look at what we call "stability." If we abandon the concept of static equilibrium, we will nevertheless have to consider a dynamic equilibrium that comprises fluctuations around one or more states of a system. If we assume a single average state with fluctuations which never depart very far from it, we say that the state is stable. But if the fluctuations are so great that they change the average state of the system, then instability exists. Obviously we must define the time scale on which we are working. If we are talking about climate, we can consider the climatic system stable on a scale of three billion years, because there has always been liquid water on the planet, and the Earth has never experienced total glaciation or total evaporation of the oceans. But on a scale of 20,000 years there is a certain degree of instability; we have moved from the Ice Age to today's climate with some occasionally very sudden changes, such as the period called the Younger Dryas about 11,000 years ago, when a cold interlude set in within a few decades and interrupted the general warming trend for almost 1,000 years. As the greenhouse effect intensifies, will we begin to experience a climate similar to the present one, or a radically different climate? And will the transition be gradual, or occur in sudden surges? Everything depends on what we call "feedbacks."

We have already given some examples of feedbacks. If, for some reason or other, the climate begins to deviate from a given state, will the changes in the various climatic processes tend to accelerate or decelerate that deviation? In the first case we have a positive feedback, and in the absence of a negative feedback on another process to slow down the change, the system will become unstable: any fluctuation will end up leading it towards a very different state. We have already mentioned one example of a positive feedback involving the following effect: When the Earth's surface begins to warm up, evaporation increases; this increases the amount of water vapor in the atmosphere, which reinforces the greenhouse effect and the warming. We have referred to the case of Venus, where the CO_2-based greenhouse effect is responsible for the hellish temperatures on its surface. Would an increased CO_2 level in the Earth's atmosphere lead to this kind of nightmare? It appears that the negative feedback of cloud cover albedo would prevent it, although it did not do so for Venus (which admittedly is closer to the Sun). Paradoxically, we must not forget the negative feedback corresponding to the effect of snowfall on albedo, which might be more important in terms of warming.

The stability of the system also depends on processes in which climate is linked to biogeochemical exchanges. Of all the possible positive feedbacks, we should first of all consider the massive releases of methane in polar regions, and of CO_2 from tropical soils, if warming does begin; these phenomena would further reinforce the greenhouse effect. When it comes to negative feedbacks, we must remember that if the increased CO_2 in the atmosphere is beneficial to plant growth—which is often the case, since CO_2 is the source of carbon for photosynthesis—this increase might be limited by the greater volume of biomass, provided the forests are protected instead of cut down. Similarly, any incipient climatic change might start to alter oceanic circulation patterns, and therefore the distribution of heat between the various latitudes and the transfer of heat and CO_2 to the deep ocean.

Before giving serious, or even fatalistic, consideration to the nightmare of a lifeless Earth, we must also remember that the climatic system has already demonstrated sufficient stability to maintain itself over the ages. Climatic changes are not always disasters, and we must recognize that model-based predictions are subject to a great deal of uncertainty. All the same, we cannot forget that uncertainties can cut both ways. If we have forgotten or underestimated certain negative feedback effects, we will have exaggerated the extent of global warming. But if there are some unsuspected positive feedbacks, the change might be much more substantial than what we are predicting today.

Shouldn't we already be able to detect some warming? This is a highly controversial point. Certain climatologists point to an increase in global average temperature since 1850; others discount this finding, making the point that it expresses primarily the fact that many meteorological stations are located in cities, and that what has been measured is a warming trend in cities as they develop. This "heat island" effect is a familiar one: over the last hundred years, large cities have become systematically warmer than their surroundings. Moreover, the temperature curve shows variations—such as the cooling trend between 1950 and 1970—which have not been explained, suggesting that all these variations might be fluctuations inherent in the present climate, and do not prove any lasting trend in the direction of global warming. American journalists tended to go overboard in associating the heat and dryness of the American summer of 1988 with the onset of global warming. Will their colleagues in France do the same for the summer of 1989? It is very important to remember that one hot summer does not prove that the Earth's entire climate is getting warmer. On the other hand, the fact that at the moment there are no clear indications of warming does not prove that the models' projections are false. Perhaps the climate actually is sensitive to CO_2, but is responding with a considerable lag because of the ocean's thermal inertia. Or perhaps the change will begin quite abruptly. If we look at what happened during the glaciations, the changes are very difficult to understand without invoking positive feedbacks, and analyses of ice cores from the Antarctic show that the quantity of CO_2 in the atmosphere at that time changed as the temperature changed. But maybe it was an initial cooling which led to a decrease in atmospheric CO_2, which subsequently favored further cooling.

The problem is therefore to understand the magnitude and speed of the climatic change in response to the "shock" constituted by a rapid enrichment of the atmosphere in CO_2. As we examine this question, we must abandon both the nostalgic dream of an absolute equilibrium, and the nightmare of a radical disequilibrium—and still never forget that Nature may still have some surprises in store for us.

III

HOW WE KNOW

WHAT WE KNOW

Before discussing the economic and political debates that have recently begun to coalesce around the future scenarios depicted by meteorological and climatological researchers, and before addressing the philosophical questions raised by these prospects, I would first like to review, more systematically than in the previous chapters, some of the revolutionary developments that have affected climatic investigation, calculation, and observation methods, giving this field of inquiry a new breadth without which these debates would have no scientific foundation. This presentation will help explain the specific working conditions and methods of the groups involved in this field today, the severe limitations to which their predictions are still subject, and the prerequisites for further expansion of our understanding, prerequisites which must be a part of the scientific policies pursued on both national and international levels. These technological revolutions, without which modern global climatology would not exist, involve analyses of the records of past climates, numerical climate modeling, and observation—most especially from space—of the Earth and our atmosphere.

REMEMBRANCE OF THINGS PAST

An understanding of past climates is essential for anyone who wants to find out what the climate of the future might be, since it gives concrete examples of climates different from those which exist today, and allows us to test our models. Technological progress in recent decades has opened up better access to the archives of past climates: not historical documents, but archives consisting of sedimentary deposits (especially on the ocean floor) and accumulations of ice on the polar caps. "Access" can be taken literally, first of all: since the 1950s, it has been possible to maintain permanent scientific stations in Antarctica, including the South Pole itself, where the United States has a base; and there are now drilling methods which can extract ice cores more than 2,000 meters [1.25 miles] long, representing more than 150,000 years of accumulation. Oceanographic research vessels also carry powerful drilling equipment capable of extracting ocean-floor sediment cores that in some cases correspond to tens of millions of years of accumulation. All of this naturally costs a great deal of money, and the funding for a presence on the Southern oceans and in Antarctica by nations such as France, Great Britain, the United States, the Soviet Union, Japan, and others, as well as Southern Hemisphere countries such as Argentina and Australia, is no doubt prompted by politics as much as by scientific considerations.

Technological innovations have also brought ways of deciphering these archives. For example, improvements in isotopic analysis have yielded vital dating methods and procedures for measuring the temperatures corresponding to past deposits ("thermometry"). The ^{14}C (carbon-14) isotope of carbon is widely used to date relatively recent deposits. This isotope is a carbon atom which, from the chemical point of view, has the same properties as ordinary carbon, but whose nucleus consists of 6 protons and 8 neutrons (while the nucleus of ordinary carbon contains 6 protons and 6 neutrons). ^{14}C is radioactive, meaning that it has a tendency to decompose after an average of 5,730 years. All the ^{14}C in the atmosphere derives from nuclear reactions between cosmic rays and nitrogen atoms, and more recently, from explosions of nuclear weapons in the atmosphere.

The ^{14}C created by these processes has a good chance of being incorporated into CO_2, and it can then enter the ocean, or be used to produce organic matter in the biosphere. If a sediment sample (or a piece of wood) contains carbon, or if a layer of ice contains organic particles or bubbles of CO_2, the proportion of ^{14}C gives us an indication of its age: this proportion represents a measurement of the time that has elapsed since carbon stopped being exchanged between the sample and the atmosphere. Obviously if the sample is too old (more than 50,000 years) it will contain practically no more ^{14}C, which limits the usefulness of this dating method to relatively recent deposits. Human activities themselves have perceptibly disrupted the proportion of ^{14}C in the atmosphere: not only with nuclear explosions, but also, and for quite some time, with the combustion of fossil fuels, which has injected into the atmosphere large volumes of CO_2 that contains very little ^{14}C, since these fossil deposits are very old.

We can also make use of annual cycles, which are recorded not only in tree rings (which is useful only for very recent periods) but also in coral reefs, layers of sediments in lakes, and layers of snow in the polar caps. Major volcanic eruptions also leave behind deposits of acid aerosols in the snow on polar caps and glaciers. For very old deposits, radioactive isotopes with long half lives such as uranium-234 (250,000 years) or potassium-40 (1.3 billion years) can be used. Fossil magnetism is also enlisted: traces of the Earth's magnetic field survive in rocks and sediments of volcanic origin, and since the magnetic field has undergone a certain number of reversals of direction over the ages (fifteen or so in the last 4 million years), reference points are left behind on the sedimentary "calendar."

We can therefore date ancient deposits; we can also measure temperatures which prevailed at that time. The basis for this "thermometry" is the fact that differences in mass between isotopes produce slight modifications in evaporation and condensation conditions and in chemical reaction rates. One commonly used indicator is the ratio between the oxygen-18 isotope and normal oxygen, or oxygen-16 (8 protons, 8 neutrons). The procedure is fairly complex. The proportion of oxygen-18 in the layers of snow deposited on the polar caps depends not only on the proportion in seawater, but also on the evaporation process, and lastly on the process which generated the snowfall. The proportion of oxygen-18 existing in seawater depends on the volume of water existing, globally, in the form of ice, since when this water was extracted from the ocean by evaporation, there was a slight tendency for the heavier isotope to be left behind. But in the ocean, the proportion of oxygen-18 that will be incorporated into the carbonates forming the shells of microorganisms (called foraminifera) which are found in ocean-floor sediments depends on the temperature of the water in which the microorganism was growing. By using oxygen-18 and effects of the same type on other isotopes, it is eventually possible to establish certain elements of a climatic history, such as temperature and ice volume. Cesare Emiliani (referred to as "Anglo-Saxon" by my ["Gallic"] colleagues, but in fact an American) was the first to apply this method, starting in 1955, on sediment cores from the floor of the Caribbean.

There are also other keys that can "unlock" the archives of ancient climates: by carefully analyzing air bubbles incorporated in ice cores taken at the Soviet Vostok station in the Antarctic, a team led by Claude Lorius from the Grenoble glaciology laboratory has detected substantial fluctuations in the concentration of CO_2, which correlate with fluctuations in temperature and ice volume during the last few glaciation cycles. Some 18,000 years ago, for example, the CO_2 concentration was 200 ppm. Similar results have been obtained by a team in Bern (Switzerland), using a core taken by the Danes and the Americans in Greenland. Shifts in vegetation zones in Europe, Africa, and America during the period following the end of the last glaciation have been determined by analyzing pollen deposits. Techniques developed in order to prospect for petroleum deposits can also be used in our investigations of the past.

PREDICTION QUANTIFIED

To understand the present and predict the future—whether next week or the year 2100—we make use of calculations, or models. Only with models it is possible to perform experiments by calculation in an attempt to identify the consequences of some particular change. Specifically, it is possible to try and predict in advance, without waiting to experience it, the result of the great global geophysical experiment that humankind has undertaken by enriching the atmosphere with CO_2.

Everyone must be aware of the enormous progress made in computer science in recent years. The amount of computing power that required a roomful (say 400 ft^2) of machinery only thirty years ago can now be slipped into a shirt pocket. It is this enormous development in our computing resources that has made possible numerical weather prediction, along with general circulation models of the atmosphere and, soon, the oceans. Of course it is possible to create a simple climatic model without using computers, for example by considering only average global models of climatic parameters and studying the changes which correspond to a modification in this or that parameter. But practical questions must be asked on a regional scale, and it is impossible to divorce what happens in a given region from what is going on elsewhere on the globe. We must therefore have models which more realistically represent

92

the complexity of the interactions among the various layers of the atmosphere and the diversity of situations on the Earth's surface. Only by using "supercomputers" is it possible to store and manipulate numerical data characterizing every region of the globe and every layer of the atmosphere; even with a resolution that is still coarse (250 kilometers [150 miles] on a side) and only ten atmospheric layers, we must still work with a hundred thousand cells.

The "realism" of these models means that interpreting calculation results begins to present difficulties similar to those encountered in interpreting observations. In addition, when it comes to climate modeling rather than numerical prediction, the only way to obtain a significant result is to repeat the calculation several times: the variability of the model must therefore studied, alongside the intrinsic variability of the climatic system. Such calculations are very expensive. It is not so much the price of the hardware: these prices, in terms of computing power, are continually decreasing over time. But the models are still too approximate to represent all the complexity of the climatic system, and every time a more powerful computer becomes available, some excellent reasons are found to increase the complexity of the calculations and make full use of the new computer's abilities. Developing programs and operating a large computing center thus requires qualified people, who are very eagerly sought after in many fields of endeavor. But computer science professionals are not always willing to accept lower remuneration just because they are working in research. This fact is fairly well understood in the United States, but not always in France.

A CLOSELY WATCHED PLANET

Meteorological and climatic science needs measurements, and technological developments since the end of World War II have multiplied the possibilities with new types of instruments and especially with new observation platforms: most important of these are the artificial satellites from which we can truly get a good look at our planet.

Instruments

Some types of instruments have barely changed at all, at least in principle. These include thermometers and barometers, or more generally those instruments which take measurements *in situ*. Measurements of this kind began more than three centuries ago, on land or on shipboard, while higher-altitude measurements required platforms (balloons and aircraft). *In situ* determinations, on land or from platforms, are also useful for many types of atmospheric composition analysis, or for studies of aerosols (water droplets, ice crystals, other particles of natural or industrial origin), in which considerable technological progress has been made.

It is remote sensing, however—in French, *télédétection*—that has experienced the greatest development, not only in platforms but also in the instruments themselves. These instruments can look at the sky (and thus at the atmosphere and clouds) from the Earth's surface, but the most interesting applications involve installing them on platforms—balloons, airplanes, and above all satellites.

Another distinction must be made between passive and active remote sensing. Passive remote sensing uses instruments which measure "natural" radiation—reflected solar radiation, emitted thermal radiation—coming from the atmosphere, clouds, or the Earth's surface. Like astronomers, we can also use telescopes; but most of the time, instead of looking up from ground-based observatories, we observe the surface and atmosphere of our planet from space. Thanks to recent progress in developing detectors sensitive to infrared radiation, clouds can now be observed both at night and in daylight, provided they are not at the same temperature as the surface below them. These developments even allow us to probe the atmosphere (passively), using spectrometric methods very similar to those used by astronomers to study the temperature and chemical composition of stars. Observations in limited spectral bands, for example those dominated by the absorption of water vapor or CO_2, yield signals which depend on the temperature or humidity of certain layers of the atmosphere. One of the difficulties of this type of measurement has to do with the fact that infrared is blocked by clouds. No matter: we can then use certain radio waves (microwaves) which pass through them; like infrared radiation, these waves are emitted naturally as a function of surface temperature.

With active instruments, researchers are liberated from the constraints of natural radiation. Sodar uses acoustic waves to study the surrounding atmosphere, somewhat like bats do. With radar, an antenna emits pulsed radio waves and detects the reflection (or "echo") of these waves from various targets. The delay between emission and echo indicates the distance to the reflector; the intensity of the return provides information about the nature of the reflector and the medium that the signal has passed through. Depending on the wavelength used, the signal is affected to varying degrees by water vapor and liquid water. The principal application of radar in meteorology is to measure—remotely, of course—the liquid water content of clouds and precipitation. In the United States, television weather reports often show not only satellite pictures, but also a mosaic of "images" obtained by a network of ground-based radars which indicate the intensity of precipitation in progress. It is a pity that such images are so rare in France. Radar has other meteorological applications, too: The Doppler effect (the frequency shift that occurs proportional to the relative velocity between the observer and the object being observed) is used to measure wind speeds, and satellite-based radar is also utilized to observe the ocean surface, ice fields, and continents.

Another even more recent active remote-sensing instrument is lidar, or laser radar. The principle is the same as with radar, except that a laser combined with a telescope is used to emit light pulses, in a very narrow beam and with a very precise wavelength. With ground-based observations, the delay in the pulse reflection is used to determine the altitude of the base of a certain cloud layer, even if the clouds are very thin ones such as cirrus clouds. Temperature information can also be obtained, and aerosol layers can be detected at various levels in the middle and upper atmosphere. The various lidar stations around the world have been used to track the development and movement of aerosol layers produced in the stratosphere by volcanic eruptions: Mount Saint Helens (United States) in 1980, El Chichón (Mexico) in 1982, and many others. Lastly, by working at certain wavelengths it is possible to study the high-altitude distribution of water vapor and ozone.

Platforms

With the development of meteorological research, and then of operational predictions, has come a realization of the need to take measurements at altitudes as well as on the ground. Back in 1648, Blaise Pascal measured the decrease with altitude in atmospheric pressure (which corresponds to the weight of air above our heads) by taking a barometer to the summit of the Puy-de-Dôme (an extinct volcano in South Central France). Since the end of the 19th century, and especially since World War I, there have been systematic efforts to investigate the atmosphere with sounding balloons, carrying payloads of thermometers, barometers, and hygrometers. Recovering the data once depended on recovering the payload itself (equipped with a parachute) with its recorded measurements, but today data are routinely transmitted in real time by radio, although of course any meteorological agency would always prefer to recover and reuse its instruments. Winds can also be measured by tracking the paths of balloons that carry radar reflectors.

Many commercial airliners supply routine measurements, relatively simple but of great value, especially over oceans, deserts, and the poles. Some aircraft have also been adapted for research, carrying—often to altitudes as high as 20,000 meters [66,000 feet]—instrument packages capable of taking *in situ* samples and measurements, along with all kinds of remote-sensing equipment: radiometers, radar, lidar, etc. Samples can thus be taken *within* the clouds, which can be observed from above or below. France recently acquired a Fokker F-27 turboprop aircraft equipped for atmospheric research and remote sensing, as well as several smaller airplanes. Some other countries (Great Britain, West Germany, the United States, and the Soviet Union) have larger aircraft, or ones that can fly higher.

All over the world, at "synoptic" times (always at noon and midnight Greenwich time, and every three hours at some stations), meteorological stations on land and on weather ships take measurements and release balloons for radio soundings. The data, transmitted to each nation's meteorological services, are then retransmitted throughout the world via the GTS, the global telecommunications system that links these different departments. Each one therefore has a synoptic view of the state of our planet's atmosphere. This principle of unrestricted, no-cost exchange of meteorological information is well established and well respected, at least in peacetime.

Regular atmospheric monitoring of course depends on the existence of networks of meteorological stations. The industrialized nations of Europe and America, Japan, Australia, and South Africa, not to mention a number of other countries such as India and China, have sufficiently dense networks on their own territory to support the system. And if what is needed is a short-term local or regional forecast, for example of tomorrow's weather, these observations may be sufficient. But as soon as the weathermen try to predict the weather for the day *after* tomorrow in France—or on the west coast of the United States—they need information about the status of the atmosphere over the ocean. But it is impossible to maintain a similarly dense network of weather ships. Commercial airliners and ships supply a great deal of information, but these measurements are made only along their particular routes, which follow a limited number of sea lanes and airways. To make forecasts a week or two in advance, or to understand what is happening on the scale of an entire season, one needs information about the entire globe. But in many Third World countries, particularly in Africa, meteorological services have almost no equipment, and the only networks that exist are very low-density and also function irregularly. Over most of the oceans, especially in the Southern Hemisphere, the network is nonexistent. Global coverage can therefore be obtained only with satellites, and they are a very recent arrival. Satellites are the basis for the work of the European Centre at Reading in England, which hopes to achieve medium-term meteorological forecasts (one to two weeks).

Satellites play at least two roles. Their importance for telecommunications is well known, and has helped establish the space industry as a flourishing commercial enterprise. But satellites also act as relays for meteorological, oceanographic, and hydrological data, allowing rapid recovery of measurements made by various kinds of automatic stations, fixed or drifting, some of which are either difficult to reach or completely inaccessible. The Argos system, for example, developed by the French Centre National d'Études Spatiales [National Space Research

Center] (CNES), can locate untethered buoys or balloons or any other platform equipped with an Argos module—a small radiotransmitter—by analyzing the Doppler shift of the frequency detected by the satellite. It then becomes possible to take all sorts of measurements, anywhere in the oceans, in some cases with buoys which submerge and then resurface; these observations are important in understanding interactions between the oceans and the atmosphere. The system can also track flights of long duration by instrumented balloons for wind research, even in the stratosphere.

Balloons, the only platforms used for "conventional" operational measurements, are not the obsolete, marginal systems that one might imagine. They can carry instruments capable of making *in situ* analyses of trace gases such as ozone, or the nitrogen oxides, CFCs, or other substances which contribute to its destruction. Such flights have played a major role in recent broad-based international efforts to study the stratosphere above the polar regions. A new type of platform called the "infrared hot-air balloon" (MIR), developed by CNES in France, has proved particularly valuable. The heat sources used by this balloon are solar radiation during the day and, at night, the thermal infrared radiation which rises from the Earth's surface and lower atmosphere. Such balloons can therefore spend extremely long periods in the stratosphere, and can even circle the globe several times. But all this would obviously be useless if there were no satellites to track the balloons' trajectories and relay their data. Every such balloon is ultimately lost: If nightfall finds an MIR above a convective system of clouds that are too high and cold, there is too little infrared radiation to sustain it; if it descends too low, it becomes a danger to air traffic and must be destroyed. Until now, flights have taken place only in the Southern Hemisphere, since border crossings in the Northern Hemisphere are generally fatal to all kinds of research balloons, for reasons which are easy to guess. We can only hope that the increasing openness of the Soviet Union will soon make it possible to perform this kind of research on the Arctic polar vortex.

Beyond their role in meteorological telecommunications, the greatest significance of satellites lies in their function as Earth observation platforms. Since the beginning of the Space Age, and in fact even before the Sputnik launch in October 1957, meteorologists and geophysicists in general had foreseen the new possibilities that artificial satellites would bring, and lost no time turning them into reality. The first American weather satellite, TIROS-1, was launched in 1960. At present there exist at least a dozen continuously operating satellites specifically dedicated to meteorological observation, and numerous others designed for research on the atmosphere, the oceans, and land surfaces, all of which present problems more or less directly linked to climatology.

Not all satellites have the same function, or the same significance for weather forecasting or climatological research. Satellites in polar orbit, meaning those which pass over the Arctic and Antarctic polar regions *and therefore over every latitude* on each orbit (approximately every 100 minutes), provide truly global coverage. Since its orbital plane is relatively fixed in space and the Earth revolves beneath it, a single satellite of this type can observe the entire world every 12 hours. Many recent satellites have been placed so their orbital plane rotates just enough to compensate for the Earth's motion around the Sun, so that every longitude is observed at approximately the same local time. These orbits are therefore called "Sun-synchronous." NASA's Nimbus series research satellites, for example, always pass overhead at noon and midnight, local time, which facilitates comparisons from one day to the next.

These satellites are generally placed in relatively low orbits (between a few hundred and a thousand kilometers), so that passive observations can be made with good spatial resolution and, if necessary, active instruments can be carried. For operational meteorology, the tools are imaging instruments—telescopes—sensitive to visible and IR radiation, capable of resolving details on the order of one kilometer. At this scale, certain aspects of clouds or surface features will obviously escape observation. These aspects can be investigated with instruments of the same type but considerably finer resolution (down to ten meters [33 feet]), such as those carried by the Landsat or SPOT satellites dedicated to Earth resources research. There is generally no need for the ultrafine resolution (less than one meter [3 feet]) which must be available for "spy" satellites.

Several years ago, scientists found a way to use the observations made by polar meteorological satellites for research which was not exclusively meteorological or climatological, but ecological. By comparing the visible-light image with the near-infrared image—these being the two images corresponding to reflected solar radiation—it was possible (in cases where there was no permanent cloud cover) to determine a "vegetation index," since healthy vegetation reflects particularly well in the near infrared. This makes it possible to monitor changes in biomass in response to the seasonal cycle and climatic anomalies, and also as a function of changes in atmospheric CO_2 levels.

In fact, the most valuable data for operational meteorology come not from images, but from other instruments, "sounders" which use infrared and microwave spectrometry methods. These sensors are used to measure changes in temperature and humidity as a function of pressure and altitude in the atmosphere all over the globe (not just where radio soundings are made), twice every 24 hours. These data are essential for weather prediction.

For the past several years, the meteorological department of NOAA (National Oceanic and Atmospheric Administration) has made an effort to keep in orbit two polar satellites—one passing overhead around 2:00 a.m. and 2:00 p.m., and the other around 7:00 in the morning and evening. The data measured by these satellites are continuously transmitted to Earth to anyone who wants to intercept them; some of the global information is stored on board and retransmitted to the United States. In France, the Space Meteorology Center (CMS) at Lannion collects the observations made by these NOAA satellites over Europe and part of the North Atlantic. The U.S. Department of Defense also operates a polar meteorological satellite (and sometimes two) for its own requirements; the information obtained by these satellites becomes available for civilian research after a certain time delay, when it is no longer usable for forecasting. The hydrometeorological service of the USSR also maintains two civilian meteorological satellites in polar orbits, under conditions similar to those of the NOAA polar platforms.

At present, the Europeans are primarily making use of data from the American polar satellites, but the European contribution to the equipment carried by these satellites is far from negligible. Some instruments were developed by the British, and France supplies and operates the Argos system which, as we have seen, is capable of localizing and collecting data from mobile platforms such as balloons and buoys. Similar modules can rapidly locate ships or aircraft in distress—this service is provided by the SARSAT (Search and Rescue Satellite) system, a cooperative undertaking by the United States, Canada, France, and the Soviet Union that uses American and Soviet polar meteorological satellites.

There can be no doubt that during the next decade, the Europeans and the Japanese will be called upon to play an even larger role in the Earth Observing System (EOS) by putting into orbit "polar platforms," in other words large polar-orbiting satellites carrying a substantial complement of instruments designed to observe the atmosphere, oceans, and land surfaces. This will reduce the risk that the system might be crippled, for budgetary reasons, by the U.S. Congress, which in recent years has suggested several times that only a single NOAA satellite be maintained.

The Europeans and the Japanese are already participants in the area of geostationary meteorological satellites. These platforms, which play an essential role in telecommunications, are satellites placed in a circular orbit above the equator at an altitude of 35,700 kilometers [22,170 miles], which corresponds to an orbital period of 24 hours. Since the Earth rotates once on its axis during this period, a satellite of this kind appears to "hover" over a given point on the equator. Its field of observation is therefore limited to one portion of the Earth; in particular, it cannot observe the polar regions. On the other hand, it is a remarkably good way of tracking processes which occur in equatorial and tropical regions up to the middle latitudes, even when they are progressing rapidly. These satellites carry telescopes that observe in both the infrared and the visible; with the infrared images, obtained every half-hour, clouds can be tracked around the clock. By monitoring cloud movements, especially over the ocean, wind speeds can be determined at various altitudes, and this information is vital for forecasting models. In some cases images can be obtained at an even more rapid rate over a limited area, in order to track phenomena—such as violent thunderstorm systems over the United States—that develop very quickly.

Since the end of 1977 (with a hiatus between November 1979 and June 1981) Europe has deployed the Meteosat series of satellites on the Greenwich meridian above the equator; in addition to the European continent, these satellites provide excellent observations of Africa, the Atlantic, and even part of South America. Ever since the early 1970s the United States has made an effort to maintain two operational geostationary meteorological satellites (GOES East and GOES West) to cover the Americas and the eastern Pacific. Several years ago the Japanese deployed their GMS or *Himawari* (sunflower) satellite to monitor the Western Pacific and East Asia. Serious difficulties have been encountered over the sector comprising the Indian Ocean and Central Asia. The Soviets are understandably more interested in polar satellites and have not yet launched a geostationary meteorological satellite, although they announced plans for one some time ago. They were even supposed to cover the Central Asian sector, to complete the geostationary satellite chain, during the first worldwide atmospheric research experiment in 1979. In the end it was a spare American GOES which was placed in that position during the experiment, with satellite control relay and data collection provided by the ground station at Lannion in France, working at the limit of visibility. More recently, a series of geostationary satellites (INSAT)—combining the roles of meteorological monitoring and television/telecommunications relay—has been put into service by India; after considerable frustration, data are now starting to become available for processing. It must be mentioned that the Americans have also encountered their share of breakdowns, launch pad accidents, and delays in producing and deploying the GOES series. At present, the Europeans are loaning a spare Meteosat to the United States to ensure observation continuity until it can be replaced.

In addition to the satellites and instruments which provide the information routinely used by operational meteorological agencies to forecast the weather, there are also those which supply data for research in all the other fields associated with climatic studies. The instruments carried by NASA's ERBS satellites monitor not only the planet's radiative balance (including solar flux), but also certain stratospheric gases and aerosols. The CZCS (Coastal Zone Color Scanner) on the Nimbus 7 satellite has been used to study ocean color, especially in regions close to land, where this color can provide information about the phytoplankton which play a major role in CO_2 exchanges between ocean and atmosphere. Measurements of this type will also be made with a Japanese satellite. Other radar-equipped satellites can measure sea level to within a few centimeters; this parameter depends on oceanic current systems, and varies substantially during the events associated with an El Niño. The French-American Topex-Poseidon project should soon allow considerable progress in this field. Various types of radar can be used to monitor the status of the ocean from space, including wave heights, from which surface wind speeds can be deduced. The ERS-1 satellite operated by the European Space Agency will be dedicated to this type of measurement. International cooperation in space is progressing: In late 1991, a Soviet Meteor satellite is scheduled to carry the American TOMS (Total Ozone Mapping Spectrometer) to measure ozone levels, and in 1993 another Meteor will carry the joint French-Soviet-German instrument called ScaRaB (Scanner for Radiation Balance) to monitor the Earth's radiative balance.

There are plans to increase even further the variety and capabilities of future Earth observation satellite payloads. One particular goal is to equip satellites with lidars in order to observe the atmosphere and cloud cover from above. The first one is to be installed on the Soviet Mir space station, within the framework of joint French-Soviet space operations. In the longer term, there are plans to use lidars on polar platforms to provide general coverage of atmospheric winds by utilizing the Doppler effect; the technological hurdles are enormous, but the potential rewards substantial.

These, in a nutshell, are the observation instruments and platforms that are revolutionizing, before our very eyes, not only meteorology and climatology, but more generally most of the Earth sciences. Technology in this field is progressing very quickly, and there can be no doubt that many of these advances in techniques for observing the Earth (and the Universe in general) would never have seen the light of day without research financed largely for military purposes. This is obviously true for the development of radar, infrared detectors, large sectors of aeronautical science, rockets, and satellites. In a future which—let us hope—will see a decline in international tensions and thus in military budgets, can we count on a continuation of the research efforts which are helping us to understand our planet better? The conquest of space seems to be well underway, but is by no means assured. We must also remember that the satellites and their instruments, both automatic and remotely controlled, which provide almost all the observations of scientific, economic, and even military interest, cost considerably less than the activities involved in manned spaceflight. The Hermès manned spaceplane project is devouring much of the funding allotted to space research in France and in Europe, but progress in scientific understanding will come primarily from the instruments installed on unmanned platforms. Although the media tend to emphasize information about manned space-flight, it is worth remembering that while the Soviet Union's ambitious space exploration program does indeed involve maintaining a practically continuous human presence in its Mir space station, with several launches a year, over the same period it launches close to a *hundred* unmanned satellites (almost two a week).

People

Amid all this celebration of technology, we must not overlook the essential role played by the men and women who work in research. All too often, space agencies are happy to put industry to work building and launching satellites, without worrying too much about what will be done with them. But at the end of the chain of instruments and computers, human intelligence is still essential. Satellites transmit an enormous flood of data: for the needs of operational weather forecasting it may be enough to process them "on the fly," but if we want to make progress, especially in understanding climate, that information must be stored and then studied as long-term series. Magnetic tapes are piling up and threatening to inundate many research laboratories. It was not until 1983, more than 20 years after launching the first weather satellite, that an international program was developed in which the various agencies began rationally and effectively sorting their incoming data and creating the summary archives which are now allowing researchers to establish the climatology of cloud cover as observed by satellite.

Theory and modeling are considered, perhaps especially in France, to be noble undertakings; but observation, particularly when it comes to that overly familiar planet that is our own Earth, seems to be somewhat neglected. But we must not make the mistake of thinking that Earth observation is a problem governed by satellites and by the automatic systems which process the data that they generate. Not that we should be content with merely qualitative descriptions of what these images show us. There is no question that the flow of data is so large that computerized processing is indispensable. But processing programs are still a human creation, and the algorithms on which they are based depend on our explicit or implicit models of nature. We must continue to examine these observations and the results of the automatic processing programs with a critical eye, which in turn must be trained by research. To paraphrase Hamlet, "There are more things in Heaven and Earth than are dreamt of in our models…"

Consider a very specific example: The discovery of the ozone hole was announced in 1985 by a British team working on the ground with "conventional" instruments and examining its observations in detail. Only later, after reexamining the data transmitted by the TOMS instrument on NASA's Nimbus 7 satellite, was it found that the hole had already been forming for several years. Why had nobody noticed it? The reason was simple: the systems processing the TOMS data, designed in accordance with predictions derived from models, which in turn were established on the basis of what was thought to be "reasonable," had rejected the very ("excessively") low values observed above the

Antarctic during the Southern spring: as far as the program was concerned, there must have been an operating defect in the instrument. Although researchers were looking for—and were measuring—a generally decreasing trend in ozone levels, they were not prepared to accept something that had not been predicted in the models. If Nature has other such surprises in store for us, will we be able to recognize them in time?

IV

CLIMATE AND THE
POLITICAL SCENE

CLIMATOLOGY AND POLITICS

Stockholm, 1972: a conference on the environment under the aegis of the United Nations; Geneva, 1979: the first world conference on climate. During these meetings, the scientific world began to talk seriously to the political world. In 1988, in order to protect the stratospheric ozone layer, 120 countries agreed to limit their use of CFCs and signed the Montreal Convention. In 1989, several heads of state and heads of government of the industrialized nations signed the Hague Appeal to alert public opinion to current threats to the environment. Over the last decade or so, governmental authorities have suddenly realized that climatic change is one of the questions on which the planet's future depends. The rising power of environmentalist movements and their growing electoral clout, especially in Western Europe, have certainly contributed to this realization. In 1979, no one felt any need to meet at the ministerial level. More than ten years later, even the British Prime Minister, Mrs. Thatcher, had discovered the significance of the environment. And after spending years laying what they called the "cornerstones" of an essentially *extra*terrestrial research program, the European Space Agency is now hastily "greening up" its few Earth observation satellites.

This new situation, which researchers have greeted with understandable elation, nevertheless exposes them to all sorts of pressure which should worry them, and which should be provoking some reaction from them. Not satisfied with calling on researchers to find answers to questions which they can just barely formulate in scientific terms, now everyone wants science to give its authoritative blessing to decisions which are, in principle, beyond its competence.

One exemplary case that is nonetheless easy to resolve is certainly that of the notorious CFCs. On the basis of preliminary research, certain countries such as the United States and Sweden had already prohibited the use of CFCs in aerosol cans towards the end of the 1970s, but most other nations refused to follow their example. It took the discovery of the ozone hole to provide impetus for the Montreal Convention in 1988, which provided that signatory nations, themselves the largest users of CFCs, had to reduce their consumption by 50% between then and 1999. The Federal Republic of Germany has announced that it will have reduced its consumption by 95% by the end of 1989; the figure for France is 90% by 1991. It will not always be that easy in certain industries, but the commitments will nevertheless be met without too much difficulty; the reason is that the search for substitutes undertaken by the chemical industry—including the American giant Du Pont—has proved successful. The Montreal decision therefore does not present any major economic problem. The fact that this sort of decision was made before the scientific questions were completely answered is not at all surprising: after all, it is the responsibility of "decision-makers" to proceed on the basis of necessarily incomplete information.

Today there is considerable political pressure—and equally strong resistance—towards proceeding similarly in the matter of CO_2 emissions. But most of the questions involving the greenhouse effect cannot simply be answered by government regulation and technical innovation, since these questions lie at the heart of the basic economic activities—agricultural production and generation of energy—of every human society. When we examine these questions even-handedly, giving consideration to every aspect, we always come up against political problems, especially the problem of what to do about the so-called "developing" countries.

But just for a moment, let us examine this example of the increase in CO_2 levels in the atmosphere. We have seen that it has essentially been attributed to deforestation in tropical regions, and to a general increase in industrial activity. Can the industrialized countries limit, let alone reduce, their emissions of CO_2? It is possible, at not too great a cost, to "scrub out" the emissions of SO_2 (also responsible for acid rain) which correspond to the small quantity of sulfur contained in coal or petroleum. On the other hand, there is apparently no way to prevent CO_2 from escaping into the atmosphere without losing all the benefit of the energy produced during combustion: Let us not forget that for each ton of coal burned, more than three tons of CO_2 are produced. Must we then stabilize, and eventually reduce, our use of fossil fuels? In the wealthy countries that is certainly a possibility, and market forces have already led to a definite improvement in energy utilization efficiency since the "oil crisis" of 1973. We must also remember that conservation measures can have their own costs: if a house is too tightly insulated in the interest of reducing heating costs, it collects radon, a naturally occurring radioactive gas. While efficiency is being improved, the search is therefore on for new energy sources which might replace fossil fuels (coal and petroleum). Some suggest imposing a tax on CO_2 emissions, so that market mechanisms will push energy producers towards conservation and substitution; others advocate the government regulation approach; still others would prefer to wait and see. These are questions for economists and politicians—but they cannot be answered without an understanding of the laws of nature. What if the alternative energy sources turn out to be nuclear power plants? Certain suspicious types have gone so far as to suggest that politicians are emphasizing the threat of the greenhouse effect in order to promote nuclear power, and it is true that a nuclear power plant does not generate any CO_2 at all.

But what about the Third World? After all, it is a fact that most of the predicted rise in CO_2 emissions is predicated on predictions of substantial economic development to be expected in Third World countries. Some of these countries, like China, have enormous reserves of coal. But must we assume, in the scenarios that are being devised, that the development of the Third World, if indeed it occurs, will proceed on the basis of Western-style industrialization exploiting the same sources of energy on the same scale? Might we not instead imagine other routes of development that lead towards other models of social well-being, or at least use different technologies that would consume less oil and coal? Do we really want to see increased use of nuclear power? Could solar energy and its exploitation turn out to be something more than a decoy (as in Pierre Boulle's novel *Miroitements*, Flammarion, 1982, about sparkling mirages)? Can we advise the Third World to aspire to something other than a consumer society without setting a suitable example ourselves? None of these, obviously, are questions that scientists can answer, even if they can give their opinion about the outcome of the policy selected, drawing up a balance sheet strictly from the point of view of changes in the composition of the atmosphere and the consequences of such changes.

The same applies to deforestation. We know that tropical forests grow on soils which are usually poor, and we also know that clearing this land and using it for grazing yields results that are always mediocre and rapidly become catastrophic. This has been proven once again in Latin America. The forest itself may prove to be a source of considerable wealth—in the pharmaceutical field, for example—which has by no means been completely inventoried. Without even envisaging the repercussions in terms of CO_2, rainwater recycling, and the climate in general, we already have good reasons to protect the forest. Turning to the use of wood for fuel, in Africa this practice leads to unspeakable waste. In the Sahel, it is not so much the Sahara that is encroaching; it is the people, in their search for wood, who are creating a desert. When a population can no longer feed itself from its land using traditional methods, the situation demands new methods. But does anyone really know what the best policy is for rationally and sustainably exploiting tropical lands in the best interest of their populations? And suppose someone did know, would international aid and financing organizations want, or be able, to not only recommend but impose these utilization policies in the face of demographic pressures and human misery, the inertia of age-old customs, and the nationalistic reflexes which are inevitably stirred up whenever the greed of special interests is impeded? What coercive measures would we then have the right, and the ability, to apply?

The case of methane is even more glaring. Increased methane levels can no doubt be attributed—according to many specialists—to increased cultivation of rice and to more widespread cattle breeding. The situation with nitrogen oxides, associated with the use of artificial fertilizers, is similar. One cannot seriously suggest, on the basis of an intensifying greenhouse effect and fears of climatic change, that entire populations radically change the way in which they produce basic foodstuffs. There is no way we can revert to the methods of cultivation of our ancestors. Undoubtedly everything would be simpler if one zero could be knocked off the world's population figure: of course the Third World countries, like the industrialized nations, must some day soon stabilize their populations, but that will not happen specifically for climatic reasons, and it will not solve the problem of our changing climate.

From a global point of view, how are we to interpret the climatic "New Deal" that is being heralded? The real difficulties, as we have already mentioned, may arise from excessively rapid changes. Waiting to get a clear picture of the changes that are now occurring may perhaps mean waiting too long to be able to change course, or even to adapt to the new situation. Politicians, already continually overtaken by events, have ended up listening to the scientists and ecologists, and recoiling at the specter of the ozone hole. But it did not cost them very much to eliminate CFCs. What will they do when faced with the threat of changes that affect the very foundation of the wealth of nations? Must we necessarily regard an increase in CO_2 as an evil in itself? There is nothing inherently noxious about CO_2: it is the raw material of plant growth and improves their water consumption efficiency. As for us humans, we routinely tolerate concentrations well above 1000 ppm in our meeting rooms. We have seen that global warming would produce increased precipitation in some areas, and reductions elsewhere. There is no justification for thinking that this increase would be catastrophic for many countries in Africa... Would international economic relationships be realigned as a result? And are they now so satisfactory that realignment would be such a threat? Were the United States to produce a little less wheat while Canada and the USSR produced a little more, is that such a nightmarish prospect? In a world where billions of human beings cannot get enough to eat, can we decently dread such a change? This climatic catastrophism seems to me to be a one-way fear, and frankly I regard it as a sign of shocking egotism. Moreover, is it not a convenient way to conceal other urgent problems?

I am firmly convinced that scientists will not be able to participate fully in these difficult debates until they learn to resist the entreaties of politicians, who are pressuring them to conjure away the major uncertainties which still bedevil their research findings. Such resistance will not be easy, because the pressure is strong. As an inevitable consequence of their function, politicians balk at taking responsibility for decisions made on uncertain foundations, especially if they entail substantial cost. Instead, they demand short-term certainties, or seize upon uncertainty as a pretext to justify their inertia. When it comes to exerting pressure, however, they have powerful resources, since they are the ones who determine research budgets. Consider, then, the enormous temptation to force results a bit in order to satisfy these expectations. This is a very serious question of responsibility, of "ethics," as we would say today. The scientist must give an accounting of his research to the community that is financing it, but he also must not conceal his uncertainties, and must refuse to give the illusion of answering questions which have no real scientific significance. Let us all, as scientists, find the courage to turn aside these questions, and force politicians to make the choices that truly exemplify the exercise of democracy. ▪

STEERING THE COURSE OF OUR PLANET

It is often said that humankind has disrupted the balance of Nature. I have already stated that this concept of balance or equilibrium, no matter how useful it might be to scientists, is still a caricature not to say fallacy when we start talking about our environment. But this myth of a balance of Nature, cultivated by the many ecological movements which are now doing such a brisk business, has the effect—and in fact may be the cause—of papering over the real, political nature of the questions being asked. The idolatry of Nature cultivated by these movements often has overtones of actual mysticism, the practical and political consequences of which can be dangerous, as we have seen in the past, when dressed in colors less cheerful than green. Like most scientists who study the Universe from the infinitely large to the infinitely small, I share the sense of wonder and even reverence at the grandeur and beauty of these natural phenomena. Like many people who have chosen research as a profession, I am perhaps moderately misanthropic, and I look with disdain and disgust on certain aspects of our overdeveloped, greedy, and bloated societies. But I do not hate the human race to the point of espousing a cult of Nature from which we would be excluded! After all, it should be obvious that we long ago transformed our environment—or, if you prefer, that we profoundly denatured it. The idea of "pure Nature," so familiar to the philosophers of the Enlightenment, is even more fallacious today than it was two hundred years ago.

Whatever may be said, ours is not the first living species to modify the environment, or to produce changes in climate. Consider algae, for instance: a very long time ago, they profoundly changed the composition of the Earth's atmosphere by adding oxygen to it. For microorganisms with a sulfur metabolism, to whom oxygen was a deadly poison, this was a dreadful act of pollution, as a result of which they can no longer exist in the open air. Algae are

still affecting our climate today because they excrete dime-
thyl sulfide, promoting condensation and thus the
formation of clouds. According to the "Gaïa hypothesis"
formulated by the English chemist James Lovelock, the
Earth is practically a living organism with certain homeo-
static mechanisms. He argues that the biosphere as a whole
acts like a climate control system which maintains opti-
mum conditions for life on Earth. But what sort of "life"
does he mean? This provocative theory becomes highly
debatable when we try to give it precise scientific content,
since life has suffered innumerable upsets and transforma-
tions over the past three billion years. If life had not
managed to recover after the catastrophes of the past, we
would not be here to talk about it. But the successes of the
past are no guarantee for the future.

What is new about humankind is not that we are trans-
forming Nature, nor even that we are doing it very fast, but
that we are capable of understanding the changes that we
are introducing. While lemmings hurl themselves blindly
into the sea, we with our science are observing changes,
modeling possible futures, and beginning to learn how to
the discern the consequences of our actions. Are we equally
capable of controlling them? In short, the real question
before us is whether we are willing to guide the course of
our planet. Because if we really can perceive where a given
policy is leading us, and if we can really modify it on the
basis of that knowledge, we can—within the limits
imposed by the laws of nature, of course—choose our own
destiny. The scientific understanding and political mecha-

nisms which may spare us from an unwelcome climatic change are the same ones which may allow us to modify the climate deliberately. What happens to climate would then be inextricably linked to the future of humanity.

Perhaps this point of view is too optimistic when it comes to the ability of human societies to control their own behavior. Will politicians be prepared to jeopardize the established order, "business as usual," on the basis of uncertain predictions about a more or less distant future?

From an economic point of view, what is the value of the future? The problems promise to be formidable, when we remember that the gains and losses associated with global warming will be unequally distributed around the world, just like coal and oil resources or scientific and technical know-how.

Can we steer the course of our planet without global dictatorship? Would battles for access to the levers of power lead to worse catastrophes than the anarchic continuation of present trends? Perhaps we are condemned to know that we are heading for catastrophe, without being able to retrace our steps. But if a catastrophe is inevitable, in terms of the relationship between human beings and their environment, it will not necessarily be the end of life on our planet, or even of our species. The Earth has seen other disasters, even though it may take thousands of years for life to begin blossoming again. It would then be up to the people of this distant future, far less numerous in an ecologically impoverished world, to "protect Nature" by reining in the natural instincts of our species to proliferate and dominate. Scant consolation for us and our children...

As a scientist, I cannot accept the idea that some ineluctable Last Judgment is imminent. Today, the protection of Nature is only one aspect of human control of Nature, which in turn must be based on our self-control; control within the bounds of fundamental laws, but control nonetheless. The question then remains: will human beings have the wisdom—or perhaps the temerity, the *hubris*—to utilize the scientific, technical, and above all political resources which will allow them to "steer" our planet? Or would this be a headlong, relentless pursuit, leading to catastrophes of even greater magnitude? Of course we cannot literally steer the planet without a clearly designated destination; but what should that destination be? The question, I repeat, is a political one in the short term, but ultimately it is also a philosophical one. I would be gratified if these few pages helped my readers to confront it.

BIBLIOGRAPHY

BERGER, A., SCHNEIDER, S., and DUPLESSY, J.-C. (eds.), *Climate and Geo-Sciences. A Challenge for Science and Society in the 21st Century.* Kluwer Academic Publishers (NATO ASI Series), 1989.

BERROIR, A., *La météorologie* [Meteorology*]. Presses Universitaires de France (*Que sais-je?* series), 1986.

BOLIN, B., DÖÖS, R., JÄGER, J., and WARRICK, R.A. (eds)., *The Greenhouse Effect, Climatic Change and Ecosystems.* SCOPE Ser.: No. 29, J. Wiley, 1986.

FELLOUS, J.-L. (summer school ed.), *Climatologie et observations spatiales* [Climatology and space-based observation*]. Cepadues Éditions, Toulouse, 1987.

"Managing Planet Earth," special issue of *Scientific American*, No. 145, November 1989.

HOUGHTON, J.T. (ed.), *The Global Climate.* Cambridge University Press, 1984.

HOURCADE, J.C., MÉGIE, G., and THEYS, J., "Politiques énergétiques et risques climatiques. Comment gérer l'incertitude?" [Energy policy and climatic risks: managing uncertainty*]. *Futuribles*, No. 135, pp. 35–60, September 1989.

KANDEL, R.S., *Earth and Cosmos*. Pergamon Press, 1980.

KONDRATYEV, K.Y., *Climate Shocks: Natural and Anthropogenic*. Wiley–Interscience, 1988.

LABEYRIE, J., *L'homme et le climat* [Man and climate*]. Denoël, 1985. "La Recherche Météo" [Meteorological research*], special supplement to *La Recherche*, No. 201, July 1988.

LE ROY LADURIE, E., *Histoire du climat depuis l'an mil* [Times of Feast, Times of Famine: A history of climate since the year 1000]. Flammarion, 1967.

LOVELOCK, J.E., *The Ages of Gaïa: A Biography of Our Living Earth*. W.W. Norton & Co., New York, 1989.

MÉGIE, G., *Ozone, l'équilibre rompu* [Ozone: the balance upset*]. Éditions du CNRS, 1989.

RAMADE, F., *Les catastrophes écologiques* [Ecological catastrophes*]. McGraw-Hill, Paris, 1987.

SCHNEIDER, S.H., and LONDER, R., *The Coevolution of Climate and Life*. Sierra Club Books, San Francisco, 1984.

---

* These references have not been published in English.